Rethinking Food System Transformation

Rachel Bezner Kerr • T. L. Pendergrast •
Bobby J. Smith II • Jeffrey Liebert
Editors

Rethinking Food System Transformation

Second Edition

Previously published in *Agriculture and Human Values* Volume 36, Issue 4, December 2019

Springer

Editors
Rachel Bezner Kerr
Department of Global Development
Cornell University
Ithaca, NY, USA

T. L. Pendergrast
Cornell Prison Education Program
and Department of Global Development
Cornell University
Ithaca, NY, USA

Bobby J. Smith II
Department of African American Studies
University of Illinois Urbana-Champaign
Urbana, IL, USA

Jeffrey Liebert
Faculty of Agricultural
and Environmental Sciences
McGill University
Montreal, QC, Canada

ISBN 978-3-031-30483-5 ISBN 978-3-031-30484-2 (eBook)
https://doi.org/10.1007/978-3-031-30484-2

This Springer imprint is published by the registered company Springer Nature Switzerland AG
The registered company address is: Gewerbestrasse 11, 6330 Cham, Switzerland

Contents

Introduction to the Symposium: Rethinking Food System Transformation—Food Sovereignty, Agroecology, Food Justice, Community Action and Scholarship 1
T. L. Pendergrast, Bobby J. Smith, Jeffrey A. Liebert, and Rachel Bezner Kerr

Food Justice, Intersectional Agriculture, and the Triple Food Movement 7
Bobby J. Smith II

Pockets of Peasantness: Small-Scale Agricultural Producers in the Central Finger Lakes Region of Upstate New York 19
Johann Strube

Action Research on Organizational Change with the Food Bank of the Southern Tier: A Regional Food Bank's Efforts to move beyond Charity 31
Alicia Swords

Gardens and Green Spaces: Placemaking and Black Entrepreneurialism in Cleveland, Ohio 49
Justine Lindemann

Participatory Plant Preeding and Social Change in the Midwestern United States: Perspectives from the Seed to Kitchen Collaborative 61
G. K. Healy and J. C. Dawson

To Save the Bees or not to Save the Bees: Honey Bee Health in the Anthropocene 73
Eleanor Andrews

Agriculture and Human Values (2019) 36:819–823
https://doi.org/10.1007/s10460-019-09952-z

SYMPOSIUM/SPECIAL ISSUE

Introduction to the symposium: rethinking food system transformation—food sovereignty, agroecology, food justice, community action and scholarship

T. L. Pendergrast[1] · Bobby J. Smith II[2] · Jeffrey A. Liebert[3] · Rachel Bezner Kerr[1]

Accepted: 22 May 2019 / Published online: 27 June 2019

Within the last decade, there has been a growing interest in merging community-based knowledge with scholarly voices to understand how food systems can be used to create or maintain sustainable, just models of social change (Alkon and Agyeman 2011; Millner 2017; Schipanski et al. 2016). As an example of such efforts, Patricia Allen's (2008) publication in this journal, "Mining for Justice in the Food System: perceptions, practices, and possibilities," marked an important scholarly call-to-action and mapped out ways academics could have an active role in shaping a more "diverse agrifood system that embraces the discourse of social justice" (p. 157). According to Allen, educators can help students explore, understand, and de-naturalize the root causes of persistent food inequalities related to production and consumption; researchers can craft participatory projects on food issues; and scholars can work to inspire improved policy and practice (Allen 2008, p. 160). Allen's hopeful piece presented issues as opportunities and systemic ills as possibilities for new strategic alliances within academia and between researchers and non-academics. In the 10 years since Allen's piece, much more scholarship on the subjects connected to food justice, food sovereignty, equity in agriculture, and agri-food system problems has emerged, and examples of creative strategies for reform are widely evident (Alkon and Agyeman 2011; Holt-Giménez and Shattuck 2011; Holt-Giménez and Wang 2011; Heynen et al. 2012; Clendenning et al. 2016).

Looking back on the last decade of scholarship and on the words of Allen (2008) in particular, we see that the authors of this symposium ask: What are different advocates, stakeholders, growers, and community members today "mining for" when it comes to justice, action, and transformation in the agri-food system? The research conducted by the scholars in this symposium reflects the diverse range of approaches scientists have taken to investigate this aforementioned question. The papers also come out of a conference, detailed below, that represented a combined effort to creatively educate, share, and connect work being done by stakeholders. Reviewing the origin story of this conference not only supplies more context for the content of the papers that follow, but this story also details how these pieces were presented to non-academics to connect research meaningfully to community.

The 2017 *Farm-to-Plate: Uniting for a Just and Sustainable Food System* conference in Ithaca, New York, was in many ways exemplary of Allen's call to "foreground and work towards social justice in the food system" (Allen 2008, p. 157). This conference grew out of visions from and exchanges between Damon Brangman, a local Ithacan farmer, staff at the Groundswell Center for Local Food and Farming, Rachel Bezner Kerr, Professor in the Department of Development Sociology and Noliwe Rooks, Professor in Africana Studies, both at Cornell University. Thus, from the beginning, the conference included a range of perspectives and leadership from community members, small-scale growers, nonprofit sector advocates, and academics. However, the subsequent planning process was by no means easy: different concepts of the conference's purpose, who should be included and how, the scope of the event, and

✉ T. L. Pendergrast
tlp59@cornell.edu

Bobby J. Smith II
Bobbyjs2@illinois.edu

Jeffrey A. Liebert
jal485@cornell.edu

Rachel Bezner Kerr
rbeznerkerr@cornell.edu

1 Department of Development Sociology, Cornell University, Ithaca, USA

2 Department of African American Studies, University of Illinois at Urbana-Champaign, Urbana, USA

3 School of Integrative Plant Science - Soil and Crop Sciences Section, Cornell University, Ithaca, USA

other aspects became subjects of long deliberation and, at times, friction. The planning committee meetings revealed the tensions within different food system-oriented entities in Ithaca and the perceived divisions between groups in town and the university across racial, gender, and class lines. The planning sessions often shifted into discussions about power, inclusion, and equity, as experienced by different committee members. The careful process of putting together the conference's logistics and elements echoed a desire to bridge divides between academics and non-academics.

Despite some frictions stemming from intersecting issues and interests, the conference was widely considered a success by the various partners who planned it as well as by the attendees. The opening night highlighted some of the key focal points of the event by bringing together a panel of Black food justice activists, including Jamila Simon, Rafael Aponte, Malik Yakini, and Karen Washington. These activists commented on a range of observations from their advocacy, for example noting that we need to ensure that food justice projects are led by the marginalized themselves and that power relations between academics and community members continue to be a point of concern warranting attention. These conversations on power relations and transformative possibilities continued at the next morning's events. Breakout sessions and workshops brought together farmers, small business owners and academics for public dialogue about the struggle to use ecological and social justice principles in their work.

The keynotes by Raj Patel and Malik Yakini amplified these aforementioned themes in the context of food sovereignty. Each speaker emphasized the need to acknowledge connections between current inequalities in the food system and past injustices in order to truly build food system transformation. Workshops, presentations, field trips, and small group discussions followed these presentations, allowing for fruitful dialogue, exchange, and learning amongst a diverse group of local community members, activists, farmers, academics, students, and non-profit organizations. Reflections from attendees highlighted the unusual mix of intellectual, practical and activist orientations that permeated the event (see https://groundswellcenter.org/farm-to-plate-conference-re-cap/for more information on the conference).

The intellectual foci of the papers in this symposium mirror these orientations and considerations of the Farm-to-Plate conference itself. They illuminate the challenges of community conversations on food system transformations and the potential of inclusive discussions on food justice to break new ground in thinking, in community action, and, importantly, in scholarship. Although several of the papers recognize the merit of community-centered models as opposed to large-scale industrial ones, the authors featured here attempt to avoid the trap of reductionism and essentialization in their focus on local communities. Instead, these papers underscore the complexity of the communities being analyzed in case studies, from the groups of people of color involved in re-imagining spaces in post-industrial landscapes to the networks of queer farmers creating new possibilities for themselves or others. In this way, each paper in this symposium contributes to a larger understanding of farm-to-plate relations at the nexus of locality, social movement activism and coalition-building, and food system participation and transformation.

Symposium contributions

The contributions of this symposium examine farm-to-plate relations as an effort to deepen our understanding of food system alternatives by exploring the narratives and models seeking to address unsustainable food production and the injustice and inequality that result. While some of papers consider the various means of fostering agroecological production methods within a social justice context, others focus on knowledge creation and the role of experts when faced with concerns about food sovereignty, food justice, and control over resources. Together, these papers explore the possibilities, connectedness, and differences between movements and practices linked to food justice, food sovereignty, and agroecology. Concepts such as *food justice* and *food sovereignty* have arisen out of social movements, both as a critique of the dominant agri-food system (Alkon 2014; Clendenning et al. 2016) and as a way forward to create a sustainable and equitable food system. Food justice, built on earlier environmental justice scholarship, examines the ways in which systemic racism and class inequalities shape access and control over healthy, sustainable food supplies with its origins and focus largely in the urban United States (Alkon and Agyeman 2011; Clendenning et al. 2016). In contrast, food sovereignty arose from rural social movements, and is more broadly focused on shifting political control of the food system away from the highly concentrated agri-food industry back into both producers and eaters' hands (Wittman et al. 2010). As an alternative approach to farming, agroecological management attempts to mimic natural systems through energy, soil, and water conservation, increasing biodiversity, and leveraging ecological relationships among crops and other organisms across spatial scales (Gliessman 2015). Although many scholars continue to describe agroecology in narrow terms, a more comprehensive understanding of the concept—one in which sociopolitical and cultural dimensions are explicit and indivisible from the ecological aspects—has increasingly been intertwined with food justice and food sovereignty approaches in more recent years (Méndez et al. 2013; Holt-Giménez and Altieri 2013). Recent scholarship has argued that *transformative agroecology* uses participatory action research that draws on local and indigenous knowledge, and addresses social inequalities within the food system (Méndez et al. 2013). Yet, there is limited research to date on bringing these different approaches—food justice, food sovereignty, and agroecology—together in spaces that integrate scholarship with action (Brent et al. 2015). Thus, an underlying goal of this symposium is to begin to address this gap, with a series of papers focused on the galvanization of local and regional food system transformation using food justice, food sovereignty, and agroecology concepts together with participatory, action-oriented, community-centered research.

Bobby J. Smith II, in "Food Justice, Intersectional Agriculture, and the Triple Food Movement," challenges another perception of similarity across a form of community, one that presupposes LGBTQ communities are predominantly affluent and food-secure (2017, p. 20, cf Brown et al. 2016, p. 2). To do this, he uses food justice documents to situate how eco-queer farming reveals overlapping vulnerabilities of poverty, discrimination, and lack of access to institutions. These concerns of LGBTQ food producers are both unique in some ways and run parallel with those of other struggles, such as black farmers addressing the food needs of low-income people and people of color in urban areas. Smith in turn considers the similarities and intersections between these two struggles in terms of approaches to activism. He sees and offers what he coins "intersectional agriculture" as needed to combat systems of domination that marginalize particular experiences. Intersectional agriculture, to Smith, maintains that "simultaneously addressing food issues and structures of inequality such as racism, classism, or sexism" is critical to agri-food system reformation.

Eleanor Andrews, in "To Save the Bees or Not Save the Bees: Honey Bee Health in the Anthropocene," explores how upstate New York beekeepers produce knowledge and make sense of what constitutes sustainability in their management of bees. She argues that the framework of the 'anthropocene,' a term for a geological age marked by human influence on environmental systems, is helpful for understanding the dilemmas beekeepers face when it comes to their work with bees. Her qualitative research explores ideas of what qualifies beekeeping practices as natural, commercialized, community-oriented, or innovative to those who work with bees. Her interviews reveal a world of entanglement: one in which a form of human–environment interaction defies simple binaries like traditional/modern and wild/domesticated. In her consideration of research like Tony Weis' (2013) work on systemic effects for species, Andrews wonders about the bee as a representative of the complexities of the Anthropocene; these creatures are managed by humans and in turn shape us tremendously. Andrews explores the data on honey bee health and the debates about population crises, but ultimately her focus takes readers to the community of bee-keepers themselves and the kinds of knowledge they draw on, refine, and re-make for a changing environment.

Justine Lindemann, in "Gardens and Green Spaces: Placemaking through Arts, Food, and Culture in Cleveland, OH," interrogates how conventional community development models have tended to "focus on singular problems such as unemployment or malnutrition" versus the interplay of needs within the arts, culture, entrepreneurship, equity, and food production. She argues that this interplay could enhance how communities make sense of place-making, rather than relying on the conventional single-focus community development models that do not adequately address broader systems of oppression. Lindemann unravels the 'participation-as-buy-in paradigm' that many community development models employ by examining what she terms 'needs-only' perspectives. Projects that address housing attainment, food, employment, and access to healthcare are all important, she argues, but many of these operate from a survivalist mentality rather than a consideration of the means to thrive. The difference between surviving and thriving might lie, in part, in types of participation: Lindemann suggests that resident-driven models of community development include participation in culturally-meaningful projects of reclaiming place.

Johann Strube's piece, "Pockets of Peasantness: small-scale agricultural producers in the central Finger Lakes region of Upstate New York," shows through qualitative research that non-market agricultural subsistence work—or 'peasant' production—is not merely an antiquated practice or a reality restricted to the Global South. Rather, peasant forms of agricultural production are visible in 'pockets' enmeshed with capitalist market forms. The documentation and exploration of these pockets of peasantness has theoretical implications. In terms of theory-building, such spaces contrast with conceptual frameworks of the dualist nature of production, either capitalist or peasant/non-capitalist. These 'pockets of pleasantness' support Van der Ploeg's (2008) suggestion that models of farming can allow for both modes and that peasantry itself is more than a population; peasantry includes associations of values and behaviors as well. Strube uses a case study of upstate New York small-scale farms to demonstrate Van der Ploeg's reconceptualization of what peasant conditions look like, providing evidence that while farmers are embedded in capitalist modes, they also grow alternatives to capitalist production alongside their market-based work. This case study gives voice to growers in upstate New York, especially those frustrated by market-related challenges, and provides stories of ways collectivity can retain vibrancy even if the capitalist model is not wholly upended.

Grace Healy and Julie Dawson's work, "Participatory plant breeding and social change in the Midwestern United

States: Perspectives from the Seed to Kitchen collaborative," focuses on the practical ways research can buttress efforts in social movements and food reform initiatives. Healy and Dawson argue that plant-breeding work is uniquely positioned to incorporate participatory project models, thereby including input from farmers and other non-academics. They describe plant breeding as applied research that necessarily needs to observe effects "on the ground," and thus this world of plant innovation is well-suited for creating participatory research models that foster inclusivity and user (e.g., farmer) feedback and guidance. The case study in this piece is the Seed to Kitchen Collaborative at the University of Wisconsin-Madison. The Seed to Kitchen Collaborative has sought to design plant breeding projects with farmers' goals in mind and keep meaningful, productive, and realistic participation for those using seeds as a primary objective. In explaining the work of the Collaborative, Healy and Dawson explore the growing global efforts to achieve and maintain 'seed sovereignty'; much of their article explores the idea of seed self-determination and the ability of communities to take ownership over the means of present and future food supplies. Healy and Dawson conclude with a celebration of the positive steps taken towards seed sovereignty and participatory plant breeding and they suggest some ways the Collaborative could enhance the project outcomes if capacity were increased.

Alicia Swords' article, "Action research on organization change with the Food Bank of the Southern Tier: a regional bank's efforts to move beyond charity," speaks to the conundrums of some food security models, particularly in upstate New York's economically-challenged landscape. Like Healy and Dawson, Swords pursues participatory methods, in this case through the engagement of action research. Her study is a story of what she describes as the 'emergency food regime'; building on Philip McMichael's (2005, 2009) analysis, Swords develops one facet of what McMichael calls the current epoch of corporate food control. The emergency food regime does, on the surface, seem to address a social ill, but the regime does not work to solve more deeply embedded inequalities. Swords notes that even without the regime analysis, food pantries and charity models have been the subject of a wide array of critique. Taking these critiques of emergency food together with theories on persistent poverty, Swords portrays the Food Bank of the Southern Tier as an example of a food relief effort that is trying to step beyond the charity model, and thus transcend some corporate food sphere limits. Swords finds there is potential for this food bank and others to use their networks, experiences, and role in their community to forward discourses on food inequality, poverty, and models of alternatives to charity fixes.

Conclusion and future directions

Although these papers examine different parts of the food system, they help illuminate the often-overlooked connections between food justice, food sovereignty, agroecology and participatory approaches. The works of this symposium highlight community efforts to remake local food systems towards greater equity, inclusion, and ecological resilience. The articles explore theoretical frameworks for understanding complexity in agri-food networks, and the authors deepen discussions of practices like participatory methods, intersectionality, and food sovereignty in community development through rich case studies and historical analysis. The ways in which these papers were developed also demonstrates the theories at work with attention to process, audience, methods, and purpose of scholarship. For instance, scholars presented their papers to diverse audiences in accessible public spaces and the feedback and questions the researchers received were notably broader in scope and more nuanced in terms of actions and community outcomes than what would be expected at a strictly academic conference. As a result, conversations generated from those presentations helped shape the papers in this symposium. The process by which these papers came to be shared, discussed, and revised is an exciting form of participation that complements reflection on the transformation action within the case studies of the articles themselves.

We suggest further research that invites productions of community-university partnerships and local knowledge-building. At a time when many universities are asking academics to be more community-centered and stakeholders are understandably holding researchers more accountable to community impact, the process by which this symposium came to be demonstrates a model for more engaged, community-based scholarship. As researchers, we should continue to reflect on the way we carry out our research activities, with whom and how we collect stories and how we share and revise our works with and within community. Pursuing inclusive and innovative approaches that place a high value on community input and the potential implications stemming from our research is also important for improving researcher-community relations. In short, we see the production of this symposium as a call to action itself—one that builds on Allen's ideas for the engagement of academics and the potential of scholarship to contribute to community-based efforts that seek justice and transformation.

Let us in academia also work collaboratively in our communities to provide more spaces for conferences that include diverse voices from non-academics and multi-form ways to connect research projects with the work of organizers and practitioners. As we write, we note that there is another Farm-to-Plate conference in Ithaca in the works; many in the community felt the impact of the inaugural event and are excited to come together at the table once more. We look forward to seeing what future iterations of this conference and others like it generate on the page and as collective, participatory, transformative action.

Acknowledgements This symposium arose out a collective event held in May 2017, called the Farm to Plate Conference, co-organized by Roots Rising Farm, Groundswell Center for Local Food and Farming, and the Food, Agroecology, Justice and Wellbeing Collective at Cornell University. The involvement of Damon Brangman, Kate Cardona, Dr. Noliwe Rooks, Dr. Bobby Smith II, and Rafael Aponte is particularly acknowledged in making this conference a success. Funding for the conference was generously provided by Engaged Cornell, Polson Institute for Global Development, Einaudi Center for International Studies, Institute for Social Sciences (all of Cornell University), the Park Foundation, Sustainable Tompkins Mini-Grant Council, and the GreenStar Natural Foods Market.

References

Alkon, Alison Hope. 2014. Food justice and the challenge to neoliberalism. *Gastronomica: The Journal of Critical Food Studies* 14 (2): 27–40.

Alkon, Alison Hope, and Julian Agyeman. 2011. *Cultivating Food Justice: Race, Class and Sustainability*. Cambridge: MIT Press.

Allen, Patricia. 2008. Mining for justice in the food system: Perceptions, practices, and possibilities. *Agriculture and Human Values* 25: 157.

Brent, Z.W., C.M. Schiavoni, and A. Alonso-Fradejas. 2015. Contextualising food sovereignty: The politics of convergence among movements in the USA. *Third World Quarterly*. 36: 3.

Brown, Taylor N.T., Adam P. Romero, and Gary J. Gates. 2016. *Food insecurity and SNAP participation in the LGBT community*. Los Angeles: The Williams Institute.

Clendenning, J., W.H. Dressler, and C. Richards. 2016. Food justice or food sovereignty? Understanding the rise of urban food movements in the USA. *Agriculture and Human Values* 33: 165.

Gliessman, Steve. 2015. A global vision for food system transformation. *Agroecology and Sustainable Food Systems*. 39: 7.

Heynen, Nik, Hilda E. Kurtz, and Amy Trauger. 2012. Food justice, hunger and the city. *Geography Compass*. 6 (5): 304–311.

Holt-Giménez, Eric, and Miguel Altieri. 2013. Agroecology, food sovereignty, and the new green revolution. *Agroecology and Sustainable Food Systems*. 37: 1.

Holt-Giménez, Eric, and Annie Shattuck. 2011. Food crises, food regime, and food movements: Rumblings of reform or tides of transformation? *The Journal of Peasant Studies*. 38: 1.

Holt-Giménez, Eric, and Yi Wang. 2011. Reform or transformation? The pivotal role of food justice in the US Food Movement. *Race/Ethnicity: Multidisciplinary Global Contexts* 5 (1): 83–102.

McMichael, P. 2005. Global development and the corporate food regime. In *New directions in the sociology of global development*, ed. F.H. Buttel and P. McMichael. Bingley: Emerald Group Publishing Limited.

McMichael, P. 2009. A food regime analysis of the 'world food crisis'. *Agriculture and Human Values* 26: 4.

Méndez, V.Ernesto, Christopher M. Bacon, and Roseann Cohen. 2013. Agroecology as a transdisciplinary, participatory, and action-oriented approach. *Agroecology and Sustainable Food Systems*. 37 (1): 3–18.

Millner, Naomi. 2017. 'The right to food is nature too': Food justice and everyday environmental expertise in the Salvadoran permaculture movement. *Local Environment, Journal of Justice and Sustainability*. 22 (6): 764–783.

Schipanski, Meagan E., Graham K. MacDonald, Rosenzweig Steven, M. Jahi Chappell, Elena M. Bennett, Rachel Bezner Kerr, Jennifer Blesh, Timothy Crews Laurie Drinkwater, Jonathan G. Lundgren, and Cassandra Schnarr. 2016. Realizing resilient food systems. *BioScience* 66 (7): 1.

Van der Ploeg, J.D. 2008. *The new peasantries. Struggles for autonomy and sustainability in an era of empire and globalization*. London: Earthscan.

Weis, Tony. 2013. *The ecological hoofprint: The global burden of industrial livestock*. New York: Zed Books.

Wittman, Hannah, Annette A. Desmarais, and Nettie Wiebe (eds.). 2010. *Food sovereignty: Reconnecting food, culture and community*. Blackpoint: Ferndale Publishing.

T. L. Pendergrast is a PhD candidate in the Department of Development Sociology at Cornell University.

Bobby J. Smith II is the Chancellor's Postdoctoral Fellow in the Department of African American Studies at University of Illinois at Urbana-Champaign.

Jeffrey A. Liebert is a PhD candidate in the Soil and Crop Sciences Section at Cornell University.

Rachel Bezner Kerr is a Professor in the Department of Development Sociology at Cornell University.

Agriculture and Human Values (2019) 36:825–835
https://doi.org/10.1007/s10460-019-09945-y

SYMPOSIUM/SPECIAL ISSUE

Food justice, intersectional agriculture, and the triple food movement

Bobby J. Smith II[1]

Accepted: 9 May 2019 / Published online: 16 May 2019

Abstract
Emerging as an intersectional response to social inequalities perpetuated by the mainstream food movement in the United States, the food justice movement is being used by marginalized communities to address their food needs. This movement relies on an emancipatory discourse, illustrated by what I term intersectional agriculture. In many respects, the mainstream food movement reflects contention between marketization (corporate agriculture) and social protectionist (local food) discourses, while the role of food justice remains somewhat unclear as it relates to the mainstream movement. Each movement attempts to restructure the ways in which food is distributed, consumed, and produced, impacting the social, cultural, political, economic, and environmental dimensions of food. Using the lens of Nancy Fraser's triple movement framework, I construct an interpretation of food justice as the emancipatory pole of what I term the triple food movement to explore the role of food justice as it relates to the mainstream movement. Specifically, I draw upon the cases of black farmers and queer people in the U.S. creating and (re)creating spaces to address their community food needs and counter systems of domination constructed around race, class, gender, sexuality, agriculture, and food.

Keywords Food justice · Corporate agriculture · Local food · Triple food movement · Intersectional agriculture

Abbreviations

BPP	Black panther party
DBCFSN	Detroit black community food security network
LGBTQ	Lesbian, gay, bisexual, transgender, or queer
USDA	United States Department of Agriculture

Introduction

Emerging as an intersectional response to social inequalities perpetuated by the mainstream food movement in the United States, the food justice movement is being used by marginalized communities to address their food needs. Even before scholars began to write about food justice, activists in low-income communities of color used the term to address the lack of healthy food options in urban neighborhoods (Alkon 2012). In many respects, the mainstream food movement reflects contention between marketization (corporate agriculture) and social protectionist (local food) discourses in a Polanyian "double movement" sense (Polanyi 1944), while the role of food justice remains somewhat unclear as it relates to the mainstream movement. Yet, each movement attempts to restructure the ways in which food is distributed, consumed, and produced, impacting the social, cultural, political, economic, and environmental dimensions of food.

According to Robert Gottlieb and Anupama Joshi (2010), food justice seeks "to achieve equity and fairness in relation to food system impacts and a different, more just, and sustainable way for food to be grown, produced, made accessible, and eaten" (p. 223). In their collection of essays, *Cultivating Food Justice,* Alkon and Julian Agyeman (2011) argue that essential to food justice is "an analysis that recognizes the food system as a racial project and problematizes the influence of race and class on the production, distribution, and consumption of food" (p. 5). New York City-based food justice organization Just Food defines food justice as the practice of "communities exercising their right to grow, sell, and eat [food that is] fresh, nutritious, affordable, culturally appropriate, and grown locally with care for the well-being of land, workers, and animals" (Alkon and

Symposium Presentation: Rethinking Food System Transformation: Food Sovereignty, Agroecology, Food Justice, Community Action, and Scholarship.

✉ Bobby J. Smith II
bobbyjs2@illinois.edu

[1] Department of African American Studies, University of Illinois at Urbana-Champaign, 1201 West Nevada Street, Urbana, IL 61801, USA

Julian Agyeman 2011, p. 6). In a recent study of food justice organizations, Hislop (2015) defines food justice as "the struggle against racism, exploitation, and oppression taking place within the food system that addresses inequality's root causes both within and beyond the food chain" (p. 24).

The Oakland, California-based People's Grocery argues that "food justice asserts that no one should live without enough food because of economic constraints or social inequalities" and offers a "different approach to a community's needs that seeks to truly advance self-reliance and social justice by placing communities in leadership of their own solutions and providing them with the tools to address the disparities within our food systems and within society at large" (Mares and Alkon 2011, p. 75). Social scientists Agyeman and McEntee (2014) posit that food justice focuses on both "outcomes and processes as well as symptoms (e.g. immediate needs such as inadequate access to food) and causes (e.g. structural inequalities) of food injustice" as well as the institutions responsible for such injustices (p. 211). However, what is common among all definitions is the contention that food justice uses agriculture and food as vehicles to address social, cultural, political, economic, and environmental relations between inadequate access to food and larger societal structures of inequality.

As a social movement, food justice emerged within the context of the 2007–2008 global "food crisis" (McMichael 2014) as a response to inequalities embedded in both the dominant corporate agriculture movement and myriad local food movements. To do this, the food justice movement relies on an emancipatory discourse that interrogates the "historical-materialist relations responsible for the creation and re-creation of unjust circumstances" (Agyeman and McEntee, 2014, p. 216) to "mobilize [activists] at the grassroots level to dismantle the classist and racist structural inequalities that are manifest in the consumption, production, and distribution of food" (Mares and Alkon 2011, p. 75). "The work of food justice activists," Christine Caruso argues, "seeks to not only address issues of access at the neighborhood level, but also has a more comprehensive political vision that incorporates systemic change at multiple levels, including the local, state, and global scales" (Caruso 2014, p. 1). Moreover, the discourse of food justice relies on what I call *intersectional agriculture* and a structural interpretation of US food movements that highlights the nexus between the historical, social, cultural, economic, political, and environmental contexts of food politics.

Relying on Kimberlè Crenshaw's (1989, 1991) concept of "intersectionality," intersectional agriculture examines how structural inequalities related to race, class, gender, and sexuality produce instances of hunger and food injustice. For example, the food justice movement draws upon intersectional stories of the state-sanctioned evictions of both Native Americans (Norgaard et al. 2011) and black farmers (Green et al. 2011) in the south that have altered both populations' relationship with land; the development of gentrification practices in urban areas that have erased access to fresh, healthy foods in black neighborhoods across the U.S. (Alkon 2012); the plight of farmworkers and their working conditions (Brown and Getz, 2011); and the struggle of queer farmers to create autonomous agri-food spaces (Sbicca 2012). More specifically, for black farmers, land historically represented a sense of both economic and food security. In the case of Native Americans, land was historically and culturally embedded in the sacred relationship between nature and humans, linked to food provision and land stewardship.

Using the lens of Nancy Fraser's (2011, 2013) triple movement analytic framework, I construct an interpretation of food justice as the emancipatory pole of what I term the triple *food* movement to explore the role of food justice as it relates to the mainstream movement. As a revised version of Polanyi's (1944) "double movement" project, Fraser's triple movement framework introduces a third project to his framework to understand contemporary social struggles. She terms this third project "emancipation," which seeks to dismantle inequalities and oppressive structures that arise in the tensions between marketization and social protection. Thus, Fraser's framework "delineates a three-sided conflict among proponents of marketization, adherents of social protection and partisans of emancipation" (Fraser 2013, p. 129).

This paper uses this conflict embedded in what Fraser describes as the triple movement—which represents in effect the triple food movement—to understand food justice in relation to corporate agriculture and local food. First, I briefly show how corporate agriculture is aligned with the marketization project and the local food movement is aligned with the social protection project to situate the context of the current U.S. food movement. Then, I argue that the food justice movement is aligned with the emancipation project by defining and showing how intersectional agriculture is used by marginalized communities to address and resist inequalities produced and reproduced by corporate agriculture and local food movements. More specifically, I draw upon case studies (Alkon 2012; Reynolds and Cohen 2016; Sbicca 2012; White 2011) at the intersection of race, class, gender, and sexuality to discuss how food justice aims to address food needs and counter systems of domination constructed around race, class, gender, sexuality, agriculture, and food.

The mainstream U.S. "food movement" context: corporate agriculture and local food

The corporate agriculture movement or global corporate food regime (McMichael 2014) is heavily implicated in the American agri-food system, and relies on large farms practicing conventional or commodity agriculture, supported by the corporate-controlled, transnational U.S. food system. Historically, the movement is synonymous with the industrial agriculture movement of the late nineteenth and early twentieth centuries, "propelled by mechanization, the increased use of chemicals (i.e., synthetic fertilizers and pesticides), and, most recently, the introduction of advanced biotechnologies" (Lyson and Guptill 2004, p. 373). Sociologist Thomas Lyson (2004) maps out the history of the American agri-food system in his book, *Civic Agriculture: Reconnecting Farm, Food, and Community*, and argues that commodity agriculture attempts to embed society in the market economy. This process of embedding is guided by the neoclassical economic paradigm, that when applied to agricultural production transforms the factors of production—land, labor, and capital—into commodities that can be traded within a global agricultural-food market, driven by large subsidized farms tied to large agribusinesses such as Monsanto, Syngenta, or Dow AgroSciences (Lyson 2004; Lyson and Guptill 2004).

As a result, this movement is situated in economic and political terms between market mechanisms, large agribusinesses, and large farms. Economically, agricultural marketization supports the production of cheap, highly processed foods that represent a market of "undifferentiated commodities" built upon single-cropping production systems and technology (Beus and Dunlap 1990; Lyson 2004). Politically, agricultural marketization creates a corporate middle class that shapes and regulates where, when, and how food is produced, distributed, and consumed (Lyson 2004). Thus, economic and political power is concentrated among large farms and corporations that convert staple crops into cheap, highly processed foods. Subsequently, the political-economic power of corporate agriculture produces a number of detrimental consequences for both humans and the environment, such as the marginalization of small-to-medium scale farmers, corporate control of the food system, the production of highly-processed cheap foods, high rates of diet-related diseases, soil degradation, and a host of other social and environmental problems (Mares and Alkon 2011).

In contrast, the local food movement has arisen as a social-ecological response to the corporate agriculture movement, "rooted in a critique of industrial agriculture as ecologically, socially, and economically destructive and advocates for the creation of sustainable and just alternatives" (Alkon 2012, p. 11). Its discourse is more aligned with what some are calling the mainstream "food movement," (Alkon 2012) which "argues that the purchase of local organic food is a "vote with your fork" for environmental protection because it shifts market demand to farms eschewing chemical pesticides and lowering transport costs" (Alkon 2012, p. 31). Hence, consumers who are able to "vote" with their fork benefit most directly from the movement. This ability to vote allows consumers to exercise their "food citizenship" (Lyson 2004) rights that enable them to support the economic viability of small farms that produce local or organic meats, vegetables, or fruits, and to have access to local, fresh foods. Food citizenship is defined as the status of a person that "not only has a stake but also a voice in how and where his or her food is produced, processed, and sold" (Lyson 2004, p. 77). Food citizens support what sociologist Thomas Lyson terms 'civic agriculture,' "which represents a sustainable alternative to the socially, economically, and environmentally destructive practices that have come to be associated with conventional agriculture" (Lyson 2004, p. 2).

Historically, the local food movement has origins in the sustainable agriculture movement of the 1980s. The sustainable agriculture movement is situated within "three distinct streams of social thought and activism—agrarian, environmental, and social justice" (Hassanein 1999, p. 4)—and relies on the support of small-to-medium scale farmers, environmentally concerned consumers, and the creation of what Allen et al. (2003) describe as new agri-food initiatives. Such initiatives can be largely understood as market strategies that "seek to construct and portray alternatives to the construction and reproduction of hegemonies of food (and agriculture) in the conventional food system" (Allen et al. 2003, p. 62). These strategies usually take the form of Community Supported Agriculture (CSA) organizations, farmers' markets, community gardens, direct marketing programs, farm stands, food policy councils, and other alternative food networks that bring together farmers and consumers across a diverse range of social sectors, including agronomic experts, retailers, chefs, consumers, and food journalists (Allen et al. 2003; Starr 2010).

However, new agri-food initiatives are typically realized in areas located in what sociologist Elijah Anderson (2015) calls "the white space" (see Slocum 2007). The "most visible" and "distinct feature" of white spaces is the "overwhelming presence of white people and their absence of black people" (Anderson 2015, p. 13). For example, in her study of farmers' markets, Julie Guthman (2011) draws attention to how the "overwhelming whiteness" of the local food movement creates a marginalizing food experience for people of color through specific practices and discourses. Moreover, these strategies tend to leave issues of social justice on the periphery, focusing on increasing the economic profit of small-to-medium scale farmers (Mares and Alkon 2011) and preserving the environment. As a result, the local food movement tends to be silent on issues of race and class, and "resonates most deeply with, white and middle-class individuals" (Alkon and Julian Agyeman 2011, p. 3).

In response to this silence, perpetuated by the predominantly affluent and white character of the local food movement and the ills of corporate agriculture, the U.S. food justice movement is being used by farmers of color, low income communities and communities of color—in both rural and urban spaces—as a vehicle to address racial and economic inequalities in the U.S. food system. From the Bay area of California to upstate New York, marginalized communities are mobilizing and organizing to develop food justice strategies such as community food projects, farmer of color networks, grassroots organizations, and other coalitions to address their food production needs. However, this movement goes beyond issues of food access to address and dismantle systems of domination constructed by race, class, gender, and sexuality within the food movement and more broadly. The agri-food system, as argued by Alkon and Julian Agyeman (2011), is "implicated in many of what Omi and Winant (1994) call *racial projects*, political and economic undertakings through which racial hierarchies are established and racialized subjectivities are created" (pp. 4–5). Consequently, the local food movement is also a racial project that fails to protect *everyone* from the ills of corporate agriculture. Hence, the food justice movement, and activists that support it, seek to emancipate marginalized communities from conditions maintained by both the corporate agriculture and local food projects.

The food justice movement

The origins of the food justice movement in the United States can be traced back to the American Black freedom struggle, situated in the context of conversations on food politics in the 1960s and early 1970s. "In fact, the long black freedom struggle," historian Mary Potorti (2014) argues, "has repeatedly underscored the cultural and political significance of food, explicitly calling attention to the interlocking structures of racism and social inequality embedded in the politics and culture of food" (p. 45). For example, in her article on the Black panther party's (BPP) community food programming, "Feeding Revolution: The Black panther party and the Politics of Food," Potorti (2014) uses the BPP free food programs as a case study to illuminate the ways in which black communities historically addressed inadequate access to food. She argues that the BPP understood that inadequate food access, food insecurity, racial inequality, and socioeconomic equality could not be treated separately in the struggle for social change and racial justice. This focus on simultaneously addressing food issues and structures of inequality such as racism, classism, and sexism, is at the forefront of food justice activism, characterized by what I term *intersectional agriculture*.

Intersectional agriculture represents the trend toward agricultural practices, food distribution, and consumption activities that explicitly seek to address, resist, or counter agri-food issues at the intersection of race, class, gender, and sexuality. The term derives from Kimberlè Crenshaw's concept of "intersectionality" (Crenshaw 1989, 1991), which speaks to the ways in which facets of identity such as race, class, gender, and sexuality "intersect in shaping the structural, political, and representational aspects" (Crenshaw 1991, p. 1244) of individual experiences, or societal structures—in this case, the food system.

Hence, the role of food justice in the context of the triple food movement is to bring intersectionality into the mainstream "food movement" discourse, which tends to be silent on issues of race, class, gender, and sexuality. By relying on intersectional agriculture, food justice simultaneously (1) celebrates the resilience and agency of marginalized communities, (2) critiques the corporate and local food movements, and (3) acknowledges the sociohistorical context in which race, sexuality, class, and gender inequalities in the triple food movement exist. Moreover, the food justice movement also works to alleviate symptoms (inequalities in food access and security), while actively working to dismantle the root causes (political, social, and economic structures) of these inequalities. This is accomplished in three emancipatory components of the movement that work in tandem each other: community sovereignty, food provisioning strategies, and alternative agri-food safe spaces. Each component is embedded in the nexus between community culture, identity, and food as a form of resistance, to create or (re)create spaces, through a social process, that connects symptoms of food inequalities to the structures that perpetuate those symptoms. Moreover, alongside this creation process, communities bridge culture, self-reliance, and community responsibility to develop strategies to address and resist oppressive characteristics of corporate agriculture and local food, as well as celebrate cultural legacies and foodways.

While these strategies are similar to marketing strategies employed by the local food movement, they are re-imagined to provide for and support marginalized communities. They tend to be located in (re)created spaces that are historically black, and are often now designated as food deserts. These include urban places like West Oakland, California or Detroit, Michigan, that are saturated with liquor stores or convenience stores filled with heavily processed foods, as well as urban or rural locations where there are no grocery stores for at least twenty miles. Moreover, these strategies are also located in lesbian, gay, bisexual, transgender, or queer (LGBTQ)-identified community spaces that place the needs of queer people at the forefront of food justice efforts. This differs from the other poles of the triple food movement, in that these efforts actively resist structures of, for example, heteronormativity, and respond to issues of local governance and food availability.

As a result, these three emancipatory components connect food justice in the U.S. context to the global food sovereignty movement which struggles to maintain "the right of peoples to healthy and culturally appropriate food produced through sustainable methods and their right to define their own food and agriculture systems" (La Via Campesina 2011). However, this link to food sovereignty is realized more in relation to access, paying special attention to *terms* of food access and *who* is determining the terms (McMichael and Morarji 2010). For the food justice movement, marginalized communities *should* be determining *their* food needs as well as any other needs that impact *their* community. By bringing race, sexuality, and class, for example, to the forefront of food discourse, low-income communities, communities of color, and LGBTQ communities use food justice to bring food into larger conversations around inequitable systems constructed around food, class, race, and sexuality. In the following sections, I draw upon the case of black farmers and queer food justice activism in the US to explore and illuminate the ways in which food justice operates at the center of all three emancipatory forces and as an intersectional response to both corporate agriculture and local food.

From West Oakland to Detroit: the case of black farmers

The plight of black farmers in the U.S. is situated in a legacy of resilience, struggle, resistance, and disenfranchisement linked to the American civil rights movement in the rural south. "In fact, black landowners," Gilbert et al. (2002) argue, "were among the first to join and support the Civil Rights Movement in the rural South" (p. 2). Black landowners also mobilized and organized to develop community-based cooperatives, such as the New Poor People's Cooperatives and the Federation of Southern Cooperatives (Green et al. 2011). The struggle of black farmers, sharecroppers, and farmworkers has a long history predating and following the civil rights movement embedded in the dual organization of the south, which facilitated White reliance on black labor and black dependence on white support. This configuration reinforced the racial project honoring white authority and black inferiority. However, coming out of Reconstruction, with slavery in the near periphery, this duality was threatened. By 1900, nearly 90% of blacks lived in the South, and 83 percent lived in rural areas, mostly working as sharecroppers or tenants on cotton farms for low wages (Hurt 2011). Although blacks were an integral part of the American agri-food system and employed conventional agricultural methods, they rarely benefitted from it. Many could not afford to buy or produce their own food and suffered from poverty, hunger, and social isolation (Chapman 1940). However, "despite these challenges," Green et al. (2011) posit, "black landownership grew in the reconstruction period" but "dwindled as more lands were consolidated into white-owned plantations" (p. 53).

From 1920 to 1997, the number of black farmers decreased from 926,000 owning 16 million acres of land to less than 20,000 owning less than 2 million acres (Gilbert et al. 2002). This dramatic decrease of black farmers and black landowners impacted the cultural, economic, and social livelihoods of black farming communities in the rural south and created conditions for a long struggle with the U.S. Department of Agriculture (USDA) (Green et al. 2011). Moreover, alongside the struggle of black farmers occurred the twentieth century industrial shift within the American agri-food system, characterized by the increase in large-scale agriculture, and the corporatization of food supply chains at the local, regional, national, and global levels, which impacted many small farmers (ibid.). As a result, oppressive race relations, loss of black land, and shifts in the U.S. agri-food landscape created the context for black farmer struggles in the late twentieth century.

In September 1997, the legacy of black farmer struggles was brought to the national forefront of agri-food conversations when almost 15,000 black farmers joined a class-action lawsuit against the USDA, known as the *Pigford v Glickman* case (Wood and Gilbert 2000). Tim Pigford, a black farmer in North Carolina, originally filed the lawsuit in August of that year, alleging county USDA officials had "systematically discriminated against black farmers for years" and "illegally denied operating and disaster loans, other credit, and benefit payments" to black farmers (Wood and Gilbert 2000, p. 60). Furthermore, the suit charged that most of the county offices were directed by local white people who improperly handled and ignored complaints of black farmers (Wood and Gilbert 2000). Upon settlement of this case, each farmer was offered $50,000 and forgiveness of debt owed by to the USDA by non-private lenders (Gilbert et al. 2002).

The West Oakland farmers market

While the number black farmers and farms is declining at significantly high rates, in terms of conventional farming, which is mostly accurately captured by the USDA's Census of Agriculture, there has in fact been a resurgence of black farmers in recent years. However, these Black farmers are mostly urban farmers who do not rely on conventional methods of production and the mono-cropping systems of some of their predecessors. "These farmers are not just growing food, either," Leah Penniman of Soul Fire Farm in upstate New York recently wrote; they "rely on survival strategies inherited from their ancestors, such as collectivism and commitment to social change. They infuse popular education, activism, and collective ownership into their work" (Penniman 2015). For example, the West Oakland Farmers' Market uses the farmers' market model as a form of food provisioning linked to the struggles of black farmers, social activism, racism, and cultural recognition. "The goal of the West Oakland farmers market" Alkon (2012) observes, "is to address two sets of circumstances resulting from institutional racism: declining numbers of African American farmers and the absence of fresh foods in low-income black communities" (p. 77). Alkon (2012) further argues that the West Oakland farmers market

> constructs and celebrates black culture as both resistance against oppression and a tool to promote healthy communities and environments. Unhealthy patterns that plague many urban black communities—including high rates of diet-related diseases and lack of infrastructure and economic development—are described as the result of institutional racism. In response, this farmers market draws on and rearticulates black foodways and farming traditions to imagine other ways of being…[I]t is concerned with providing access to environmental benefits—healthy food and public space—through the creation of a local food system (p. 77).

Here, the West Oakland farmers market case connects both the structural interpretation of the U.S. food system and community sovereignty situated within the food justice discourse. Also evident is that the social, economic, political, and health issues of this community are a direct result of institutional racism, and this farmers market is being used as a tool to *emancipate* black folks of West Oakland from the ills of racism. This attention to racism and community infrastructure allows activists to connect unjust symptoms of the food system to the structures and institutions that perpetuate inequality. Moreover, this market goes beyond food to provide a space for black Oaklanders to gather and create or reclaim their rights via a participatory food system, a system in which they can be "citizens" and "vote" with their forks as a form of community sovereignty. This is different from the food citizenship of the local food movement in that community sovereignty in the food justice context is not driven by price but rather by the need and desires of a community.

The West Oakland farmers market also provides an example of how the food justice movement uses intersectional agriculture to convert marketing strategies into food provisioning strategies by linking race and sustainable agriculture. However, the market cannot be viewed in isolation from the larger context of black economic development, similar to the work of Malik Yakini and the Detroit black community food security network (DBCFSN) as well as the National Black Food and Justice Alliance. While the Network is over two thousand miles away from West Oakland, it is also linked to the struggle and resilience of southern black farmers, community sovereignty, social activism and organizing.

The Detroit black community food security network

In her article "D-Town Farm: African American Resistance to Food Insecurity and the Transformation of Detroit," sociologist Monica White (2011) investigates how the DBCFSN uses urban farming as a strategy to support black farmers, meet the needs of mostly low-income black communities, and create a community space to positively impact community health. "D-Town activists have not only appropriated public space for the purposes of creating a healthy, well-fed, well-educated, and inspired African American community," White argues, "but they also have created a sustainable community food system that fosters a sense of self-determination and self-sufficiency" (p. 415). The DBCFSN has a larger agenda that seeks

> to end relationships of dependency and educate the community about the importance of providing for themselves. In response to the failure of the local government to provide a safe community and a range of social services, D-Town farmers have worked to build community and place the earth in the center of their struggle for social transformation. Their efforts are not invested in opposing existing power structures through protest, but rather directed at contributing to the development of a safe space through the transformation of their physical environment. In this way, the Detroit Black Community Food Security Network and the D-Town Farm create the foundations for a new world marked by new ways of being (p. 415).

While White suggests that their efforts do not necessarily oppose existing power structures through protest, I argue that the existence of the DBCFSN is simultaneously opposing power structures and transforming the physical environment as a form of protest. For example, by creating a community space for black people by black people, and focusing on self-reliance, their work exposes the lack of concern for the food environments of low-income people of color in Detroit by the American agri-food system and the emerging alternative agriculture movement. "Detroiters long have had insufficient access to grocery stores" White (2011) observes, and "the problem culminated in 2007 when Farmer Jack, the last major grocery store chain serving the city, closed its doors (White 2011, p. 407). Thus, the DBFSCN protests the American agri-food system by refusing to rely on the system to provide food for their community through an analysis that posits food insecurity as a structural problem.

Hence, I argue that their remedy for this problem represents an instance of intersectional agriculture, illustrated by their reliance on community food sovereignty through "the work in food production as a strategy to demonstrate self-determination toward political and economic liberation" (White 2011, p. 411). "Most people involved in the community food-security are young White people...well connected with other White people who are doing this work and have the resources," DBCFSN founder Malik Yakini argues,

> ...they end up having a degree of control over urban agriculture in the city of Detroit—control which is inordinate to their actual numbers in the population, and that is a problem. It is an imbalance, in the city of Detroit, which is at least 80% Black people, for them to position themselves in the forefront of this movement. I'm all for cordial, cooperative relations with anyone doing this work. I'm not for any ethnic group coming in the African-American community to control any aspect of our lives, and that includes issues of food security. (qtd in White 2011, p. 411)

Yakini draws attention to the predominantly white character of alternative agri-food movements, and suggests that black communities have agency and ability to address their own issues, including food issues, through urban agriculture. To do this, the network developed the Ujamaa Food Co-op to "provide an alternative to expensive health-food stores, supermarkets, and other retail outlets, which no longer operate within the city of Detroit" (White 2011, p. 411). They also employ a concept of urban agriculture aligned with veteran food justice organization *Just Food*, relying on urban agriculture to positon their farm"(a) as a community center, (b) as a vehicle to articulate culturally relevant language about healthy food and healthy lifestyles, and (c) as a tangible model of collective work, self-reliance, and political agency" (White 2011, p. 412).

Moreover, the DBCFSN also engages in political conversations surrounding food issues in Detroit. For instance, from 2006 to 2008, the DBCFSN was appointed by the Neighborhood and Community Service Standing Committee of the Detroit City Council, to lead efforts to develop the Detroit Food Policy Council. Members of the DBCFSN "gathered existing food policies in other cities, conducted hearings and listening sessions in the community" and generated "recommendations for alternative food systems such as urban agriculture, creating citizen education guidelines, and producing an emergency response plan in the event of a natural disaster" (White 2011, p. 411). As a result, their efforts led to the establishment of the Food Policy Council in Detroit that gathers metrics and data on hunger and malnutrition on the city and makes recommendations to the Detroit City Council (ibid.).

In sum, from West Oakland to Detroit, food is being used by black communities as a tool to promote emancipation intertwined with issues of access. Black communities in these areas and others are attempting to define food access on *their* terms to support their broader goals associated with community sovereignty. D-Town Farm is also another example of a food provisioning strategy that seeks to decide *who* provides the food, *who* receives the food, and *how* the food is being distributed, with community sovereignty at the forefront. This case also illustrates how food justice discourse employs a type of intersectional agriculture that links a structural interpretation of the U.S. food system to a community's desire for sovereignty, and to their strategies for seeking food provision. Moreover, this link shows how issues of race and class shape community food experiences and illuminates the power of food within communities that have been marginalized by corporate agriculture as well as local food, and seeks to reflect the emancipatory actions of activists on the food justice side of the triple food movement.

Queer food justice activism in the eco-queer movement

Over the last 50 years in the U.S., there has been an increase in social movement activism that seeks to address the marginalization of individuals or communities that identify as lesbian, gay, bisexual, transgender, or queer (LGBTQ). These instances of activism include, for example, most recently the fight for marriage equality and other rights that challenge "dominant constructions of masculinity and femininity, homophobia, and the primacy of the gendered heterosexual nuclear family (heteronormativity)" (Bernstein 2002, p. 536). Moreover, this activism extends beyond just the fight for social equality, but also the fight for spaces that cultivate relations between food, the environment, and sexuality. For instance, sociologist Joshua Sbicca draws attention to queer activists who are fighting for such spaces under the guise of the Eco-Queer Movement. "The eco-queer movement" Sbicaa contends, "entails a loose knit, often decentralized set of political and social activists identifying as gay, lesbian, bisexual, transgender, queer or an ally of these groups, challenging binary notions of sexuality and ecology, while simultaneously transforming material and symbolic space(s) into more just, autonomous, sustainable forms" (Sbicca 2012, p. 34). The movement bridges LGBTQ, social, and environmental visions of activism and "demonstrates an understanding of the ways in which sexual relations organize and influence both the material world of nature and our perceptions, experiences, and constitutions of the world" (ibid.).

In many conversations on food access and food security at the local, national, and global levels, food issues among LGBTQ communities are often neglected, further exacerbating food inequalities among this population. In 2016, a report by the Williams Institute at the UCLA School of Law found that approximately 27% (2.2 million) of LGBTQ adults or their families were food insecure due to economic constraints and 14% of LGBTQ adults "reported running out of food for their families and not having money for more" within a 30-days period (Brown et al. 2016, p. 2). "Contrary to popular stereotypes of the LGBT community as affluent," Brown et al. (2016) argue, "research demonstrates not only widespread economic diversity…but also that LGB people are often more likely to be poor than straight people and that transgender individuals face extremely high rates of poverty" (p. 2).

Queer youth programming at bushwick campus farm

In response to the food struggles and obstacles to earning livelihoods of LGBTQ communities, queer youth and adults from California to New York City are thinking of ways to (re)create urban and rural autonomous spaces for LGBTQ people to practice what Sbicca (2012) calls "queer ecological politics of food," which focuses on the ways in which the relationship between ecology and sexuality impacts "the personal, local, organizational, and national power struggles driven by fundamental concerns of who can eat what and under what conditions" (Potorti 2015, p. 5). The ultimate goal of such spaces is to allow LGBTQ people to "collectively experience their sexuality while simultaneously striving to create more democratic, just, and sometimes sustainable alternatives to white hetero-patriarchal norms" (Sbicca 2012, p. 38). In relation to food spaces, the Bushwick Farm campus works with the Make the Road New organization and the high school Gay Straight Alliance chapter, to create a safe space for queer youth to experience their sexuality while working toward the development of a more just, sustainable NYC food system (Reynolds and Cohen 2016). To get the idea of how this farm creates this space, I draw on Kristin Reynolds and Nevin Cohen's (2016) description of farm activities and reactions of the students by Maggie Cheney, a white, queer identified farmer, in their book *Beyond the Kale*:

> She [Cheney] recounted several ways that youth in the program reacted to it: "[They'll say,] 'Oh, let's have a rainbow-colored garden box' and 'Why don't we do our GSA outside, and farm?' and 'What if we go to this conference and do a cooking demo, and be gay and cooking farm food?' […] Cheney has also used farm activities to challenge heteronormativity and to help all of the students she works with, regardless of their sexual orientation, to reconsider traditional gender roles…and that "agriculture is a perfect example of something that demonstrates all different types of gender roles—construction, cooking, sowing seed, digging in the dirt. You all these things that you could pinpoint, and say, 'Oh that's stereotypically male, and that's stereotypically female.' And then you're given this beautiful opportunity to discuss that, and try to have students understand that they don't have to be either of those boxes if they don't want to be." (p. 54)

It can be seen that Cheney uses her teaching as a form of activism (hooks, 1994) to emancipate her students from the constraints of sexual identity while simultaneously creating a more inclusive, autonomous food space for queer youth, supporting tenets of the eco-queer movement.

Queer farmers in Northern California

In 2013, queer filmmaker Jonah Mossberg, released his documentary project, *Out Here*, which explored the lives, histories, and experiences of queer farmers in the U.S. According to the documentary's website, the film asks the questions "what does it mean to be a queer farmer, is agriculture a safe space for queer people, and what are the relationships between food production and queerness?" to "give voice and visibility to queer people in agriculture and inspire a flagrant national discussion about gender and sexuality as they are related to our food system."

As a part of this project, the Fancyland land project is located north of San Francisco on twelve acres in the California county of Humboldt. "Inspired by social justice, feminist, and anti-authoritarian principles," Sbicca (2012) notes, the Fancyland project creates "queer and radical communities and individuals by being a small-scale rural resource in the following ways: acting as a site to plug into homestead projects; providing a feminist environment for learning and sharing useful rural living skills such as alternative building, appropriate technology, gardening, and land stewardship" (p. 47). According to their website, Fancyland strives to build and grow a "vibrant rural landscape by and for queer and transgender people and allies who share a common goal of dismantling oppressions, addressing privilege and keeping up the fight." To do this, Fancyland provides an *emancipatory* space for queer people and allies to mobilize, organize and strategize ways to fight against systems of domination at the intersection of sexuality, gender, food, and agriculture.

Queer food justice activism

The eco-queer movement creates a site for both queer youth and adults to link gender, sexuality, environmental, and environmental justice activism. Moreover, as Sbicca has argued, the movement can also be used to discuss intersections of sexuality, food, and the environment as well. Whether it is queer youth in New York or queer adults in California, queer people are mobilizing and using intersectional agriculture to develop spaces where they can emancipate themselves from dominant ways of being which marginalize both their sexual identities. "It can be seen that food," Sbicca contends, "provides an adhesive by which queers can develop community, challenge heteronormativity, and create sustainable alternatives to capitalist modes of industrialized agriculture" (2012, p. 45). For example, the Bushwick Campus Farm and Fancyland both resist the corporatization of agriculture by choosing to practice an alternative way of producing. Moreover, they both also challenge the dominant narrative that most farmers who participate in the local food or sustainable agriculture movement are white, heterosexual, middle-class men, by publicly being and farming. These actions foster and maintain spaces that seek to emancipate queer folks from dominant systems.

Conclusion

In exploring the role of food justice as the emancipatory pole of the triple food movement in the U.S., I have argued against treating the food justice movement in isolation from corporate agriculture and local food, a common practice that constrains our understanding of the role of food justice in the U.S. Many studies on food justice and alternative food movements tend to use the term "food movement" to describe alternative trends to the production, consumption, and distribution of food in the U.S. However, I argue, this conceptualization fails to delineate the differences between each food movement and does not necessarily capture the conflicting interests between corporate agriculture, local food, and food justice. That is not to suggest that the three movements do not intersect, but rather that the interests of the movements tend to be silenced when they are all put under the banner of a mainstream "food movement." For instance, a sole focus on corporate agriculture ignores the detrimental consequences of the movement's marketization of food on both low-income people of color and the environment. A sole focus on local food ignores the ways in which social protection efforts, which attempt to counter the detrimental consequences of corporate agriculture, marginalize and exclude low-income communities of color.

In addition, a sole focus on the emancipatory character of food justice ignores how systems of inequality, such as racism and classism, influences the social, cultural, political, economic and environmental dimensions of food. Instead, I have argued that food justice is better understood, historically in relation to the mainstream "food movement" in the US and that it relies on intersectional agriculture to address, resist, or counter the consequences of corporate agriculture's marketization and the exclusionary actions of the social protectionist local food movement. However, further research is needed to enhance our understanding of food justice and intersectional agriculture in the triple food movement. This framework could help inform future food studies, especially those that seek to parse through the food justice discourse. For instance, future research could simultaneously explore the complexity of all three food movements in the U.S. and the relationship between conventional, sustainable, and intersectional agriculture. Another area of inquiry could explore more concrete examples of how the forces of food marketization, social protection, and emancipation work together and against each other outside of the US context.

While this paper has presented two case studies to illuminate the role of food justice in the context of the triple food movement, it has focused more on issues of race and class—and less on issues of gender and sexuality—as seen through the more detailed case of black farmers. For example, I used the cases of black farmers and queer farmers to show how these two groups create and (re)create emancipatory spaces for farming, food, and community to counter the dominant U.S. food movement narrative, which posits that only white, heterosexual men farm. Yet the case of black famers shed more light on how African-Americans have been historically marginalized by food movements. In contrast, the case of queer farmers focuses more on queer bodies in spaces of food production and less on how LGBTQ communities have been excluded from the mainstream food movement. This is due to the fact that very minimal research has been done on the relationship between sexuality, food access and food justice. Moreover, queer communities have received minimal coverage in the food justice literature. As a result, we know very little about what food justice looks like in these communities and there is need for more research in this area. I propose that the concept of intersectional agriculture could be used to amplify our understanding of LGBTQ food experiences and how systems of oppression in the food system marginalize them and how they seek to resist them.

There is also a critical need for scholars and movement activists alike to expand the food justice analysis to include more explicitly issues of gender and sexuality. Although, I have argued that food justice is necessarily intersectional, most food justice projects do not engage in or practice intersectionality. However, if the movement plans to build a more equitable, sustainable food system, activists will have to expand beyond a focus on race and class and engage in intersectional agriculture to ensure that all marginalized communities are included. In the U.S., black people and LGBTQ people are two of the most vulnerable populations, therefore regularly have to develop their own 'safe' spaces. Spaces where they can empower one another and recognize their own agency in building towards a more sustainable, just food system.

Working toward intersectional agriculture could provide an avenue for the food justice movement to better align with the food sovereignty movement to address issues of gender more explicitly and the eco-queer movement in the context of sexuality. In this way, food justice provides a way for the struggles of seemingly disparate groups to be linked together in the fight against capitalistic systems of inequality perpetuated by racism, patriarchy, and heteronormativity. Interestingly, linking black, gender, and LGBTQ struggles, is exactly what Black panther PARTY co-founder, Huey P. Newton, suggested that the party do to fight against inequality in his August 1970 "Letter from Huey to the Revolutionary Brothers and Sisters about the Women's Liberation and Gay Liberation Movements" (Newton et al. 1976). "We must try to form a coalition with the gay liberation and women's liberation groups," Newton urged the party,"[w]e must always handle social forces in the most appropriate manner" (Teal 1971, p. 171). Thus, the food justice movement could be an important vehicle to which all marginalized communities come together to not only fight for access to adequate food but also issues of inequality beyond the politics of food.

Acknowledgements I would like to sincerely thank Dr. Philip McMichael, the Food, Agroecology, Justice, and Well-Being Collective at Cornell University, and the anonymous reviewers for comments and suggestions on earlier drafts of this manuscript.

References

Agyeman, J., and J. McEntee. 2014. Moving the field of food justice forward through the lens of urban political ecology. *Geography Compass* 8 (3): 211–220.

Alkon, A. 2012. *Black, White, and Green: Farmers Markets, Race, and the Green Economy*. Georgia: University of Georgia Press.

Alkon, A., and J. Julian Agyeman. 2011. *Cultivating Food Justice: Race, Class, and Sustainability*. Cambridge: MIT Press.

Allen, P., M. FitzSimmons, M. Goodman, and K. Warner. 2003. Shifting plates in the agrifood landscape: The tectonics of alternative agrifood initiatives in California. *Journal of Rural Studies* 19 (1): 61–75.

Anderson, E. 2015. The white space. *Sociology of Race and Ethnicity* 1 (1): 10–21.

Bernstein, M. 2002. Identities and politics. *Social Science History* 26 (03): 531–581.

Beus, C.E., and R.E. Dunlap. 1990. Conventional versus alternative agriculture: The paradigmatic roots of the debate. *Rural Sociology* 55 (4): 590–616.

Brown, S., and C. Getz. 2011. Farmworker food insecurity and the production of hunger in California. In *Cultivating Food Justice: Race, Class, and sustainability*, ed. A. Alkon and J. Agyeman, 121–146. Cambridge: MIT Press.

Brown, Taylor N.T., A.P. Romero, and G.J. Gates. 2016. *Food Insecurity and SNAP Participation in the LGBT Community*. Los Angeles: The Williams Institute.

Caruso, C.C. 2014. Searching for food (Justice): Understanding access in an under-served food environment in New York City. *Journal of Critical Thought and Praxis* 3 (1): 2–3.

Chapman, Oscar James. 1940. A Historical Study of Negro Land-Grant Colleges in Relationship with Their Social, Economic, Political, And Educational Backgrounds and a Program for Their Improvement. PhD dissertation, Department of Educational Studies. Columbus, OH: Ohio State University.

Crenshaw, K. 1989. Demarginalizing the intersection of race and sex: A black feminist critique of antidiscrimination doctrine, feminist theory and antiracist politics. *University of Chicago Legal Forum* 140: 139–167.

Crenshaw, K. 1991. Mapping the margins: Intersectionality, identity politics, and violence against women of color. *Stanford Law Review* 43 (6): 1241–1299.

Fraser, N. 2011. *Marketization, Social Protection, Emancipation: Toward a Neo-Polanyian Conception of Capitalist Crisis*, 137–157. In Business as usual: The roots of the global financial meltdown.

Fraser, N. 2013. A triple movement? Parsing the politics of crisis after Polanyi. *New Left Review* 81: 119–132.

Gilbert, J., G. Sharp, and M. Sindy Felin. 2002. The loss and persistence of black-owned farms and farmland: A review of the research literature and its implications. *Southern Rural Sociology* 18 (2): 1–30.

Gottlieb, R., and A. Joshi. 2010. *Food Justice*. Cambridge: MIT Press.

Green, J., E. Green, and A. Kleiner. 2011. From the past to the present: Agricultural development and black farmers in the American south. In *Cultivating Food Justice: Race, Class, and Sustainability*, ed. A. Alkon and J. Agyeman, 45–64. Cambridge: MIT Press.

Guthman, J. 2011. If they only knew: The unbearable whiteness of alternative food. In *Cultivating Food Justice: Race, Class and Sustainability*, ed. A. Alkon and J. Agyeman, 263–281. Cambridge: MIT Press.

Hassanein, N. 1999. *Changing the Way America Farms: Knowledge and Community in the Sustainable Agriculture Movement*. Nebraska: University of Nebraska Press.

Hislop, Rasheed. 2015. Reaping Equity: A Survey of Food Justice Organizations in the U.S.A. Thesis, Department of Plant Sciences. Davis, CA: University of California, Davis.

Hooks, B. 1994. *Teaching to Transgress: Education as the Practice of Freedom*. London: Routledge.

Hurt, R.Douglas. 2011. *Introduction. African American Life in the Rural South, 1900–1950*. Missouri: University of Missouri Press.

Lyson, T.A. 2004. *Civic Agriculture: Reconnecting Farm, Food, and Community*. Medford: Tufts University Press.

Lyson, T., and A. Guptill. 2004. Commodity agriculture, civic agriculture and the future of US farming. *Rural Sociology* 69 (3): 370–385.

Mares, T.M., and A. Alkon. 2011. Mapping the food movement: Addressing inequality and neoliberalism. *Environment and Society: Advances in Research* 2 (1): 68–86.

McMichael, P. 2014. Historicizing food sovereignty. *Journal of Peasant Studies* 41 (6): 933–957.

McMichael, P., and K. Morarji. 2010. *Development and Its Discontents. Contesting Development: Critical Struggles for Social Change*, 233–242. Oxon: Routledge.

Newton, H.P., B. Seale, E. Cleaver, and G. Louis Heath. 1976. *The Black Panther leaders speak: Huey P Newton, Bobby Seale, Eldridge Cleaver and company speak out through the Black Panther Party's official newspaper*. Metuchen, N.J.: Scarecrow Press.

Norgaard, K.M., R. Reed, and C. Van Horn. 2011. A continuing legacy: Institutional racism, hunger and nutritional justice on the Klamath. In *Cultivating Food Justice: Race, Class, and Sustainability*, ed. A. Alkon and J. Agyeman, 23–46. Cambridge: MIT Press.

Omi, M., and H. Winant. 1994. *Racial Formation in the United States: From the 1960s to the 1990s*. New York City: Routledge.

Penniman, Leah. 2015. Radical Farmers Use Fresh Food to Fight Racial Injustice and the New Jim Crow. *YES! Magazine*. http://www.yesmagazine.org/peace-justice/radical-farmers-use-fresh-food-fight-racial-injustice-black-lives-matter. Accessed 18 January 2017.

Polanyi, K. 1944. *The Great Transformation: The Political and Economic Origins of Our Time*. Boston: Beacon Press.

Potorti, M. 2014. Feeding revolution: The black panther party and the politics of food. *Radical Teacher* 98: 43–51.

Potorti, Mary. 2015. *Food for Freedom: The Black Freedom Struggle and the Politics of Food* (Doctoral Dissertation, Boston University.

Reynolds, K., and N. Cohen. 2016. *Beyond the Kale: Urban Agriculture and Social Justice Activism in New York City*. Georgia: University of Georgia Press.
Sbicca, J. 2012. Eco-queer movement (s): Challenging heteronormative space through (re) imagining nature and food. *European Journal of Ecopsychology* 3 (1): 33–52.
Slocum, R. 2007. Whiteness, space and alternative food practice. *Geoforum* 38 (3): 520–533.
Starr, A. 2010. Local food: A social movement? *Cultural Studies? Critical Methodologies* 10 (6): 479–490.
Teal, D. 1971. *The Gay Militants*. New York: Stein and Day.
Via Campesina. 2011. Organization: The International Peasant's Voice. Via Campesina archive. http://viacampesina.org/en/index.php/organisation-mainmenu-44. Accessed May 23, 2016.
White, M. 2011. Environmental reviews & case studies: d-town farm: African American resistance to food Insecurity and the transformation of Detroit. *Environmental Practice* 13 (04): 406–417.
Wood, S.D., and J. Gilbert. 2000. Returning African American farmers to the land: Recent trends and a policy rationale. *The Review of Black Political Economy* 27 (4): 43–64.

Publisher's Note Springer Nature remains neutral with regard to jurisdictional claims in published maps and institutional affiliations.

Dr. Bobby J. Smith II is a sociologist and currently a Chancellor's Postdoctoral Fellow in the Department of African American Studies at the University of Illinois at Urbana-Champaign. His research focuses on the relationship between food justice, agriculture, race, and inequality in historical and contemporary contexts. Dr. Smith graduated *summa cum laude* with a Bachelor of Science degree in Agriculture, with a focus on Agricultural economics from Prairie View A&M University in 2011, and he earned a Master of Science degree in Agricultural Economics in 2013 and a Ph.D. in Development Sociology in 2018 from Cornell University.

Agriculture and Human Values (2019) 36:837–848
https://doi.org/10.1007/s10460-019-09944-z

SYMPOSIUM/SPECIAL ISSUE

Pockets of peasantness: small-scale agricultural producers in the Central Finger Lakes region of upstate New York

Johann Strube[1]

Accepted: 7 May 2019 / Published online: 16 May 2019

Abstract

Some farmers in the Central Finger Lakes Region of New York balance their production between principles of peasant farming and capitalist farming. They struggle to extend their sphere of autonomy and subsistence production, while extended commodity production is often a response to external forces of the state and capital. This struggle, together with a quantitative increase of small farms, can be described as an instance of repeasantization. Based on inductive, empirical qualitative social research, this case study describes the economy and social organization of six farms in the area under investigation and explores the applicability of the rich theory of peasant farming to agriculture in this community. Besides selling commodities to pay for many farming inputs and consumer goods, the farms produce for their subsistence and that of their community. They exchange products and services with other farms, they build networks of mutual provisioning, support, and mentorship, and try to take good care of the land. This paper shows that subsistence production and peasant culture are not restricted to the past or the Global South, but also exist as pockets of peasantness on six New York farms. The perspective applied in this paper suggests that principles of peasant farming may shape farming on other US farms, too, if we accept that these principles intersect with constrains of the larger capitalist market society in which they are embedded.

Keywords Peasant farming · Subsistence · New York · Repeasantization · Local agriculture

Introduction

The term peasant is not commonly used in the context of present-day US agriculture. With the exception of Nelson and Stock's (2018) work on farming in Kansas, the term has either fallen out of fashion as an analytical tool or it is assumed that the peasantry has ceased to exist under the conditions of hegemonic capitalism (c.f. Bernstein 2010). Yet, does this assumption hold empirically? Based on interviews and participatory observations in the farming community around Ithaca, New York, this exploratory case study indicates that there are farmers who strive towards a life that is built on values such as reciprocity, solidarity, co-sufficiency, and sustainability—elements of what has been described as peasant economy. These farmers navigate their production between principles of both peasant farming and capitalist farming. The former often align more closely to their intrinsic values, whereas the latter are often a reaction to off-farm forces. The productive process is an arena of social struggle, a way to extend autonomy and to align one's economy more closely with one's own ideals. This striving towards more peasant-like production and a quantitative increase of small farmers in the area under investigation signifies an instance of repeasantization (Van der Ploeg 2008) in the Central Finger Lakes.

My interest in this research evolved in 2011 during my first visit to the United States. In New York State, I visited a number of small farms that in many ways resembled small operations from Central Europe that I was familiar with. In Austria, for example, it is normal to talk about such operations as peasant farms. In fact, there is a sense of pride that goes along with the term. I was perplexed to learn that in North America, the idea of the existence of peasant farms in the US is almost unthinkable, if not offensive, even though the farms that I saw were so similar to Austrian peasant farms. I decided to probe deeper into this phenomenon by exploring how the rich theory of peasant farming may apply in the context of one farming community in the United

✉ Johann Strube
jcs80@psu.edu

1 Department of Agricultural Economics, Sociology, and Education, The Pennsylvania State University, 111B Armsby Building, University Park, PA 16802, USA

States and why this perspective might be valuable for future research.

Peasants in the United States?

Following classical theory of development, the search for a peasant culture in the present day United States seems like a futile enterprise. Both socialist (e.g. Marx 1977) and pro-capitalist (e.g. Rostow 1960) economists long predicted the disappearance of the peasantry under capitalist relations of production. Notwithstanding, scholars noticed the persistence of the peasant farm under capitalism, including the United States. Kautsky (1988) suggested that this may be due to the particular functions that peasant farms fulfill for capitalist agriculture. Bonanno (1987) argued that the persistence of peasant farms is necessary to legitimate the capitalist transition of US agriculture as a societally beneficial project. Others have advanced that agriculture is unsuited for capitalist accumulation due to natural conditions, such as the slow growth of plants that limits the maximization of output (Mann and Dickinson 1978) or the spatial inseparability of planting, harvesting, and plant and soil management (Reinhardt 1988). Reinhardt and Barlett (1989) add that small family farms benefit from economic advantages of diversified farming such as the use of by-products, superior soil management, and less dependence on market prices of only a few commodities, while larger operations suffer from diseconomies of scale like increasing difficulty to process micro-spatial ecological information and problems in supervising labor.

Friedmann (1978) described agriculture as a dualism of capitalist agriculture on the one hand, and peasant (or small-scale) agriculture on the other, with each of these sectors having a unique economic rationale. While acknowledging the existence of this dualism, Bonanno (1987) rejected Friedmann's position that these two sectors are completely separate. Rather, at least in countries of Western Europe and the United States, he understood these two sectors as a dialectic unit of capitalist agriculture made up of two different sub-units. More recently, Lyson et al. (2008) renewed this observation of U.S. agriculture consisting of two separate sectors, small-scale and large-scale operations, with "farmers of the middle" disappearing steadily.

The question of how to define a peasant complicated the study of rural society from the beginning. As capital has penetrated family farms, these can no longer be equated with peasant farms, as for example Chayanov (1966) did. Van der Ploeg (2008) offers a solution to this dilemma. Instead of categorizing individual operations as peasant or capitalist farms, he suggests a model that allows for the existence of different modes of agriculture on the same farm. He distinguishes between three ideal typical constellations of agriculture: corporate, entrepreneurial, and peasant agriculture. The merit of this framework lies in his theory of the connections and interrelations between these three constellations. It offers a way to comprehend the reality of farmers who do not fully match any of these ideal types. His framework builds on the observation that the three constellations of agriculture represent both ideal types and specific conditions. Depending on the particularities of a farm, each of these conditions may affect farmers who have to adjust their operation accordingly.

Following van der Ploeg (2008), the peasantry is not defined by a consistent group of people, but by a set of values, behavior, and circumstances that make up the peasant condition. These include a striving towards autonomy, co-production[1] of the life on the farm as well as its resource base (subsistence production), cooperation, and often pluriactivity within "a context characterized by dependency relations, marginalization, and deprivation" (23). It is in this sense that I engage with the peasant condition, or *peasantness,* throughout this paper.[2] Repeasantization, then, describes the transition towards more peasant farming, either on existing farms or through the establishment of new ones.

The vocabulary of peasant studies has only sparingly been applied to the United States. There is a wide-spread assumption that under the hegemonic capitalist economy, the peasantry as a social class ceased to exist and differentiated into workers, petty commodity producers, or small-scale capitalist farmers (Bernstein 2010). Although structural and ideological differences between small and large, expansion-oriented farms are acknowledged (e.g. Bonanno 1987; Lyson et al. 2008), the former are rarely studied as peasant or peasant-like operations (see Nelson and Stock (2018) for a notable exception). Outside of academia, the term peasant is usually met with bafflement. None of the farmers that participated in this study self-identified as a peasant. For many the term is derogatory and connotes poverty and underdevelopment. Not wanting to misrepresent study participants, I therefore refrain from categorizing farmers in the US as peasants regardless of their mode of production. However, I describe practices that correspond to theories of peasant economies as manifestations of a peasant culture. Nelson and Stock (2018) elegantly work around this issue by using the word "peasant-like" to describe practices that are similar to those more unambiguously described as peasant without applying that term directly to farmers in the United States. I use both words interchangeably.

[1] or consubstantiality, as Gray (1998) calls it.

[2] Note that there are many more definitions of peasants or peasantries used in different contexts. See Edelman (2013) for a comprehensive review.

Subsistence production within the substantive economy

Polanyi (1944) argued that only in capitalist market societies, individuals can be assumed to act according to the neoclassical principles of rational choice decision-making and self-interested utility maximization. For other societies, Polanyi proposed, we need to employ a broader definition that understands the economy as the "instituted process of interaction between man and his environment, which results in a continuous supply of want satisfying material means" (1957, p. 248). Halperin (1977, p. 2) later shortened this substantive view of the economy as the "material-means-provisioning process" within society. Thus, if we follow Chayanov's (1966) argument that peasants organize their economy around other principles than neoclassical theory would suggest, we must integrate the social institutions and culture in which the economy of farmers is embedded.

One way of engaging this substantive analysis of the economy is by paying close attention to the subsistence practices that usually do not feature in official reports on farming. Subsistence is understood as everything necessary for the reproduction of a life that is culturally appropriate and meaningful to those who live it (Bennholdt-Thomsen and Mies 1999). It includes the material basis such as food and shelter (material subsistence), things one derives direct pleasure from (cultural subsistence), and attention and sympathy to satisfy social needs, such as love and nurture (social subsistence). To satisfy these needs, subsistence needs to be produced through labor. Thus, subsistence production is the indispensable undercurrent that connects the different forms of social and economic organization that have existed throughout history.

Bennholdt-Thomsen (2006) shows that subsistence production both requires cooperation and builds community. Individuals of a community need to meet in order to exchange goods and services for their subsistence, a process that Boyle (2012) calls co-sufficiency. Arguing from the substantive understanding that these immaterial values are in fact economic, Bennholdt-Thomsen and Mies (1999) call such networks of cooperation subsistence markets which continue to exist alongside the capitalist market.

Intrinsic and extrinsic conditions of farming

Both intrinsic and extrinsic conditions impact the activities on a farm, and farmers constantly negotiate between the two. The intrinsic condition designates the necessities that result from the farmers' own reproductive needs, values and aspirations. The extrinsic condition, in contrast, comprises the demands that off-farm actors put on the farm (such as servicing debt, taxes, and affording access to commodified land and other means of production. Although these conditions do not constitute a hard dichotomy—after all, values and aspirations are shaped by socialization within the same society that produces the external condition—they are still conceptually separate. The ability to realize one's production according to the intrinsic condition depends on the power differential between the farmer and other actors in the agricultural sector.

The relationship between the two conditions can be described as a struggle. Van der Ploeg (2008) regards the striving for autonomy, the struggle to align one's economy with the intrinsic condition of farming, as a core element of the peasant condition and conceptualizes it as a form of farmer resistance. In line with Scott's (1985) argument of the "weapons of the weak, Van der Ploeg (2008) maintains that peasants do not necessarily have to be collectively organized to be counter-hegemonic. It is through their everyday labor towards autonomy and subsistence that peasants and repeasantizing farmers directly resist their integration in the capitalist economy.

Farming in the Central Finger Lakes Region in upstate New York

The farms visited for this thesis are situated in the Central Finger Lakes of Upstate New York. Spread over three counties—Seneca, Schuyler and Tompkins—they share a common urban center, Ithaca.

The area's economy is dominated by tourism and Cornell University, a large land-grant institution. The rural areas are characterized by farms, cropland, pastures, and forest, 40.2% of the land is considered farmland (USDA 2012). Dairy farming dominates economically, while produce and grapes are also common. Only 4% of the farms are certified organic. Beside growing to sell commodities, many farmers engage in agricultural subsistence production for their own and community use. Table 1 provides an overview over selected properties of agriculture in the three counties of this study.

The area around Ithaca is known for its support of local agriculture. The city's cooperative, seller-owned farmers' market is popular with locals and tourists. Several farms offer CSA subscriptions to customers. A fifth of all operations are engaged in direct marketing (USDA 2012).

The hilly terrain and affluent urban core with opportunities for direct marketing provide favorable conditions for a community of small farms. Regardless, the decline of farming as a livelihood did not spare this region. As in other parts of the country, many farms have either

Table 1 Overview of selected properties of agriculture in Seneca, Schuyler and Tompkins Counties (USDA 2012)

Economics	
# of farms	1535
Median farm size in acres	79
% of farms by size	
0.4–4 ha	8.50
4–20 ha	26.90
20–28 ha	11.10
28–40 ha	11.70
40–56 ha	10.00
57–72 ha	6.60
73–89 ha	5.20
89–105 ha	3.80
105–202 ha	8.70
202–404 ha	4.00
More than 405 ha	3.30
Total sales in $	230,789,000
Average annual sales by farm in $	145,858
% of farms by annual sales	
Less than $2500	28.7
$2500–$250,000	59.1
More than $250,000	12.2
% of sideline farms	
Demographics	
Average age of operator	55.7
% of farms with operator older than 65 years	26.5
Gender of operator	
Female	19
Male	81
Labor	
# of people working on farms (paid or unpaid)	6165
% of farms without hired labor	68.6
% of farms with hired labor	31.4
% of farms that hired migrant workers	20.3
# of farmworkers	2484

increased in size or discontinued their operations in the past decades. Between 1945 and 2012, the number of farms in these three counties decreased by 64.6%, from 4334 to 1535 farms. In line with national trends (Lyson et al. 2008), this shift has affected medium-sized farms between 20 ha and 105 ha the most. Their share of all operations has decreased, while farms greater than 105 ha and farms smaller than 20 ha have increased relative to the total number of farms (U.S. CENSUS BUREAU 1945; USDA 2007, 2012).

In recent years, however, this trend has reversed, as new farms have started operation. In total, the number of farms in the three counties has increased by 40 between 2007 and 2012 (+2.7%). This change is uneven for the different scales of farms. The number of small farms with less than 40 ha has increased most, with 87 new farms (+10.8%) in that category. In particular, the smallest farms with less than 4 ha, which are most likely to be new farms (as opposed to farms that have changed in size and thus moved into another size range) have increased most significantly in numbers, with 37 new farms. There has also been an increase of farms in the 73–105-ha range, while farms of other dimensions have decreased in numbers (ibid.).

Methodology and research design

Understanding action as a result of both intrinsic and extrinsic conditions implies that it is insufficient to look at the doings of farmers alone to deduce their values and motivations. I therefore chose a qualitative, exploratory case study design to highlight the voice of the farmers themselves. The purpose of this study is to describe one specific farming community under the lens of the subsistence perspective (Bennholdt-Thomsen and Mies 1999) and explore the applicability of the theory of peasant farming in the context of the present-day United States. Given the shortage of research on this subject, an exploratory and descriptive qualitative case study design is merited (Creswell 2013; Yin 2018). The small sample size allows the collection of ample detail to generate new hypotheses and questions on which future research can build (Babbie 2016).

Sharing my positionality as a researcher who had worked with Central European peasant farmers allowed me to ask questions that otherwise would have been considered foolish or trivial. The disclosure of my positionality thus became an important research tool.[3] Telling informants in the field that I was looking for peasant farming in the United States, they almost always met me with perplexity. But after sharing that I had visited operations in Central Europe that were similar to theirs but were considered peasant farms, they became curious and interested to participate. Generally speaking, my etic perspective as a foreigner also helped me to question everyday behavior because everything I observed was potentially new and worth looking into (Agar 1996) and my positive experience with peasant farming in Central Europe allowed me to apply the theory of peasant farming with more ease then if I was educated in a context in which the term peasant is disregarded.

Choosing the Central Finger Lakes as my research site was a matter of convenience. I was living in Ithaca, NY as a visiting student at Cornell University. Since it was my

[3] Likewise, I emphasize my positionality and personal learning story throughout this paper as a means for you, reader, to situate the interpretation of my findings.

Table 2 Overview of farms in the study

	Farm	Description	Products for subsistence
1	Garden for self-sufficiency (7 ha)	This operation is a family home garden geared towards self-sufficiency with vegetable beds, an orchard and chickens on 7 ha. It is run by a young woman in cooperation with her retired father. The operator grew up on this land, which her parents used to keep a house garden and as pasture for racehorses. Supported through a day job in the service sector, the garden provides food and a beautiful environment for her family, one WWOOFer and guest-house tenants. The operator specifically stated that she identifies as a gardener and not a farmer	Vegetables Greens Fruit Eggs Herbs Preserves (canning)
2	Community-driven vegetable farm (12 ha)	This farm is a diverse vegetable producer, run by two young brothers in addition to their family and six WWOOF volunteers. On 4 ha (8 ha are left fallow), they grow a large number of vegetables and fruit for their own consumption, as well as for sale through direct marketing. The brothers, who did not grow up farming, bought the land from their parents who relocated to this location about twenty years ago	Vegetables Greens Fruit Herbs
3	Large-scale, organic grain operation (486 ha)	This farm is a commercial, organic grain farm on 486 ha. Besides growing a number of different crops themselves, they buy, process and market grains from other producers. They hired one farmworker full-time throughout the year and another one halftime in the summer. The farm was started in 1993 by the current operator who did not grow up farming but had worked on several farms before. The woman farmer joined him fourteen years later.	Eggs Fruit Herbs Firewood Honey
4	Organic vegetable farm (19 ha)	This farm is a commercial, organic vegetable farm on 19 ha. The operator was raised on a farm, but branched out to start his own on his parent's land. He hired a group of young men from the neighboring villages. His wife is responsible for the preparation and processing of farm products for home-consumption and raises their two sons, besides working full-time off the farm	Vegetables Greens Pork
5	Start-up vegetable farm (2 ha)	This farm is a small, diverse vegetable producer. Operated by a young couple who did not grow up farming, the 2-ha farm started in 2011, but has changed its location from its original founding. While still backed by off-farm income, the farm is growing in an effort to operate on a regular basis	Vegetables Greens Preserves (salsas, pickles etc.)
6	Organic dairy farm (931 ha)	With 700 cattle, this family farm is one of the Northeast's biggest organic dairy operations. The specialized farm grows feed and keeps their cows on 931 ha with two different farmsteads. Besides selling the milk to a major dairy in Upstate New York, they run a small farm store where raw milk, fruits and products from other surrounding farms is sold. The farm was started in 1946 and is currently in the process of being transferred to the third generation. The commercial operation is run by the operator and his two adult sons and 16 hired full-time workers	Raw milk Raw cheese Yoghurt Eggs Fruit

intention to explore the possibility of a peasant culture in the United States, I could have chosen any community as a starting point. I collected data between April and September 2015 using participant observation and in-depth interviewing. I visited six farms (see Table 2) and conducted interviews with farmers (usually the main operator, though in one instance I interviewed the farming couple together), farmworkers, and volunteers. The interviews were semi-structured, following a guide that allowed for exploration of other themes as they emerged. Apart from questions regarding basic data such as farm size, number of employees, and annual sales, the questions were open-ended. Each interview took between 60 and 90 min and was audiotaped and transcribed. One interview was conducted by email on request of the interviewee (but I still visited the farm). The respondents selected the interview locations. On four of the farms, I participated in farming activities.[4]

Beyond the farm visits, I repeatedly visited the local farmers' markets where some of the participants sold their products. In addition, I attended community events, volunteered at Ithaca's food cooperative Green Star, visited other farms and homesteads, and talked to other farmers, agricultural researchers, local food and farm activists, and representatives of various agricultural organizations. Throughout the research, I took extensive fieldnotes of my observations.

In order to account for the diversity within the field, I purposely selected the participating farms. I used maximal variation sampling (c.f. Patton 2002) to create a heterogeneous sample to cover different kinds of agricultural production. In order to challenge externally imposed definitions of farming, I included a farm that was too small to appear in the U.S. Census of Agriculture and one large-scale farm that did not conform with any stereotypical image of a peasant farm. I selected farms with different production targets and farms that were operated by men and women. In contrast to Nelson and Stock (2018), I do not treat corporate, entrepreneurial, and peasant agriculture as categorical sampling strata, since I am particularly interested in the relationship between different economic rationales on the same farm.

I settled on six farms (see Table 2) that covered these criteria and were manageable to study given the length of my stay in the field. I recruited the research participants through snowball sampling. That is, I asked informants in the field, including farmers I had already talked to, if they knew farms that would fit my requirements. Since the sampling was not meant to be representative, the selection bias of this method is acceptable.

To analyze my data, I tagged excerpts from the interview transcripts and fieldnotes that either relate to the theoretical framework or issues that were frequently brought up in several interviews. I then coded these tags with theory-driven categories (such as "sharing equipment", "social gatherings", "teaching"). I compiled all excerpts with the same code and ordered them by similarities and differences. The different codes were subsumed to 11 themes (for example, the aforementioned codes were summarized under the theme "community and cooperation"). To interpret these themes, I ordered their codes and excerpts semantically and discussed their meaning in that context. This order formed the outline for my narrative in the empirical section. Three of these themes are discussed below.

Self-sufficiency and co-sufficiency

The visited farms engage in production for self-sufficiency along with commercial production, albeit to varying degrees (see Table 2).

Although farm 6 focuses on commercial dairy production, it produces a wide array of products for home-consumption.

> It's definitely nice to go to your own store and get your own stuff. I know where the milk is and I know it's probably better quality then what's in the store. We don't go grocery shopping much. We produce milk, meat, eggs, cheese, yogurt. Usually the wintertime is when you have to go to the store because there's no produce here. I can't say no produce because there's carrots here right until the spring. My brother grows the vegetables. (farmer at farm 6)

Farms 2, 4, and 5 do not separate subsistence production from gainful production. They keep as much as they need for home-consumption and sell the rest of their yield as commodities. The operator at farm 5 doubted that she would grow vegetables if she did not eat them herself:

> I love to cook. I love to eat them. I think if I didn't love vegetables, I don't think I would have as much of an interest in farming. I don't really like to grow things that I can't eat as much. (farmer at farm 5)

Garden 1 is kept entirely for the subsistence of the family. Its yield, a wide range of vegetables, herbs, fruit and eggs, is only used for home-consumption. The gardener linked her desire for beauty and fresh food with what she defines as the good life:

> I think it is a quest for beauty. I want to be in a beautiful place and I want to eat food that is as fresh as possible, still warm from the sun. Which I guess is what I think is the good life. Just trying to live as good as possible. (gardener at garden 1)

Farm 3 follows a different strategy to secure the subsistence of the farming family. Its operation is mainly geared towards gainful commodity production and the money earned is used to buy consumer goods. Yet they also produce an array of products for themselves. They trade some of their products or services for food from neighbors and friends.

> ... with the amount of trading that we do with other farmers, time is better spent for me focusing trying to make the farm work and have the farm produce income to buy those things and other products that I can trade for other food products that we need. (operator at farm 3)

Eating local food produced by themselves, friends, and neighbors is a priority for the interviewed farmers. While

[4] At the other two farms, I was not able to participate due to my lack of training with specialized equipment.

many research participants regarded self-sufficiency as important, they stressed the fact that they would rather meet their needs through community with other local producers. Most interviewees pointed out that they do not have to produce everything themselves because they know people who provide them with other local and homemade products.

> To provide many people with one service and allow many people to provide you with what they have can be just as effective if not even more conducive to a functioning life than entire self-sufficiency. (volunteer at farm 2)

In summary, all participating farms engaged in subsistence production for self-sufficiency and co-sufficiency and considered it important and meaningful. While the largest farms of this study separated production for home-consumption from their commercial production, farms 2, 4 and 5 did not. For farm 1, products for home-consumption are the main production goal.

Community and cooperation

The cooperative effort of mutual provisioning creates community and strong social ties that would not exist otherwise. One farm volunteer explained:

> Knowing other people that can help support you can be a very humbling thing. It helps creating relationships that are beyond just social. It strengthens social relationships in a big way. (volunteer at farm 2)

This sentiment describes the network of mutual support and provisioning that exists among the farming community around Ithaca. In this network, friendships and economic partnerships often coincide and one type of social relationship facilitates the other. Farmers build a supportive community by helping one another, which includes exchange of materials, products, labor, equipment, and knowledge. While for each of these areas anonymous market solutions exist, the farmers prefer to meet their needs through arrangements with people they know.

> [...] trading with other farms that produce something that you don't produce. And you produce something that they don't produce and you both need what each other got. You could do a straight up cash thing but a lot of times...these people are friends of mine, first of all. We love to have fresh vegetables. We don't have the time to have a garden. What a great thing to be able to trade cover crop seeds with them or whatever that might be for a vegetable CSA share or some meat. That system feels really good to me. (operator at farm 3)

This network of economic relationships outside the conventional, money-based marketplace can be described as an informal trading economy. Other transactions among farms involve the exchange of money, but are guided by the same values to meet each other's needs in a supportive and reciprocal relationship.

> I've got another farm where I rent one of their fields on their farm in exchange. They don't charge me any rent but they called me last night, they are building a straw bale house, they need straw bales. I said normally they are four bucks but I sell it to you for half price because you give me my thing and then I give them cover crop seeds for their vegetable farm and they let me use their land for free. (operator at farm 3)

Beyond trading farm products, most farms support each other by sharing equipment and labor. In some instances, expensive machinery was jointly purchased and maintained by two farms. Other farmers lend their equipment for free or for some service in return. The exchange of labor usually occurs in times of peak demand, for example when a crop needs to be harvested before an approaching heavy weather. Farms 1, 2, and 5 depend on the labor of volunteers (either through the international farm volunteering network WWOOF or through direct arrangements in the local community), usually youth with interest in agriculture but little experience. Commonly, the volunteers do not get paid but receive farm products and training.

Another type of cooperation is the exchange of knowledge and experience. Usually, information about farming is shared without the expectation of reciprocity. Most interviewees praised the readiness of farmers in the community to provide technical know-how and guidance.

> That's one thing about the organic community of farmers, you can just call them up and they'll tell you how they do things. (farmer at farm 6)

Beside informal, peer-to-peer exchanges of experience, there are more organized systems in place to facilitate the distribution of knowledge. These range from regular events, such as dinners or potlucks where farmers meet to socialize and share their experiences, to self-organized, grassroots and civic organizations like New York Certified Organic (NYCO) or the Groundswell Center for Local Food and Farming, to official organizations likes Cornell Cooperative Extension (CCE) and the Northeast Organic Farming Association (NOFA).

In contrast with the capitalist market economy, the network of mutual support and provisioning observed on the visited farms is built on reciprocity and is embedded in a community based on personal relationships. It is an example of a subsistence market as described earlier. The distinction between cooperation, social activity, and economy is often arbitrary. Within the observed farming community of the Central Finger Lakes, the economy is often a cooperative, social activity for the mutual best of each participant.

However, the farming community around Ithaca is not free from competition. There seems to be a perceived necessity to be profitable to stay in business and the failure of other farms is seen as an opportunity.

> Basically in farming everyone's out for themselves. I don't think farmers want to see other farmers go out, but if they do, then it's an opportunity for them. (farmer at farm 6)
>
> We will be forced to compete with other small farmers instead of the multinationals. (farmer at farm 2)

The operator of farm 5, in contrast, has not been worried that competition with other farms will be an issue:

> [...] you always wonder about growing your competition [when teaching beginning farmers], but I think it is more important that people eat good food. We have found our niche and farms tend to find their niches. (operator at farm 5)

Subsistence-oriented farming in a capitalist world

Although all visited farms were involved in subsistence production, none of them has built a livelihood solely on these activities. All interviewed farmers perform gainful work to buy commodities for their subsistence. This can be an off-farm job or production of agricultural commodities for sale.

The motivations that drive the participating farmers suggest that subsistence is their primary production target. Good food, being outside, the living off the land, feeding others, making a living, independence, self-sufficiency, and working with a partner were mentioned as reasons to farm. Each of these drivers constitute an aspect of subsistence. None of the interviewees regarded farm growth, accumulation of capital, or financial wealth as a motivation.

The dependency on monetary income is the main reason many of the interviewees deemed subsistence farming impossible in the US today. Farmers need money to pay taxes, land, inputs, equipment, and other commodities.

> A peasant farmer to me is someone who farms on a very small, almost subsistence-like scale. In my mind, I see somebody who is very poor in a material way but not so in a spiritual or family way. Not as tied and locked into the whole capitalist financial system that I find myself completely bound in. (operator at farm 3)

Some farmers are tied into surplus production by the loans they have taken out to finance parts of their operation, exemplified by farms 3, 4, and 6 in this study. Debt forces the farms to produce more than what they need for their own subsistence to service loans and interest.

The prevailing logic of capital accumulation puts other stresses on agricultural producers as well. The farmer at farm 5 explained that it would be much easier for them to live off the land if they did not have to compete with operations from distant, yet more productive areas.

This is not to say that growth-oriented and industrial agriculture is only disadvantageous to smaller farms. For instance, some of the participating operations use the manure from industrial, organic poultry farms as fertilizer. For the operator of farm 3, this is an ambiguous situation:

> I get all of my fertilizer from these giant, industrial chicken farms. I can say that I think it's horrible, but I can also say that that's being two-faced because I buy their chicken manure and spread it on my fields. It is a good way to get a concentrated amount of manure for someone like me who is relying on manure. But that manure is coming out of this really horrendous farming system that I actually despise.
>
> So, I really feel at odds with that whole thing but also don't know what else to do. I don't have an alternative. (operator at farm 3)

The operator of farm 3 revealed internal conflicts. While being proud about his achievements, he appreciated the idea of subsistence farming and has been concerned about the sustainability of his operation.

> We are locked into the system. It's a commercial organic farm. It has its really great points and it has a lot of environmental benefits, but it's also not a system that I can say I am a 100% behind either. (operator at farm 3)

Subsistence-orientation as social struggle

The farms that participated in this study are, to different degrees, oriented towards subsistence, despite being integrated in the capitalist framework of the society at large. The observations revealed that the intrinsic condition of their production is geared towards their own subsistence, whereas extrinsic conditions force them to sell and buy commodities and make some profit.

Subsistence-orientation within the hegemonic capitalist economy cannot mean pure subsistence production. It means to strive towards a life that is more strongly connected to subsistence, which entails struggle against the institutions that seek to take advantage of farmers for their interests. For example, by building a network of mutual support and provisioning, the interviewed farmers reduce their dependence on commodities. When they try to be good stewards of the land, they perform what Van der Ploeg (2008) calls the "co-production of men and living nature", which include production for consumption and reproduction. He sees this development of agriculture and resource base as a fundamental element of the emancipation of the peasantry.

The limitations in subsistence production that the farmers in this study mentioned are not a matter of free choice, but results of extrinsic conditions of farming, such as rent, debt and tax payments, the state's protection of unevenly distributed private property on means of production, and various legal regulations and standards. Often, past management decisions, voluntary or not, will determine the scope in which farmers can operate for decades to come. If, for example, a farm borrows money to invest in one technology, they will have to continue using this technology until at least their initial investment plus interest are being fully paid back. They, thus become locked in (Tisdell 1991; Wilson and Tisdell 2001) their mode of production with little leeway to create more space for subsistence production.

Thc focus on subsistence in the self-controlled sphere of production, as well as the struggle to extend that sphere, indicate that the interviewed farmers are oriented towards subsistence, despite contradictions imposed by extrinsic conditions.

Repeasantization in the Central Finger Lakes

Repeasantization, among other trends, shapes agriculture in the Central Finger Lakes Region. Against pressures to industrialize, repeasantization manifests as a struggle. The most basic manifestation of this struggle is the improvement of a self-controlled resource base that provides the foundation for autonomy. The participating farmers approached this in two ways: The first is to add new self-controlled means of production to the farm. This allows them to be used for one's own subsistence needs and exclude them from the logic of the capitalistic market. This is the route taken by farm 5, which had bought new land to scale up its operation. Even though this purchase may not be an end in itself, it is still a form of accumulation that may inhibit other farmers from realizing their subsistence needs. This suggests, that in some cases one person's repeasantization can be another person's exclusion or depeasantization. The second way is to enhance the quality of the resource base already held by a farm. Farm 3, for example, plans to extensify its production and focus on cover crops to regenerate and build up the soil. Farm 2 plans to extensify much of their land to be able to use less land more productively. The gardener also focuses on the improvement of her land instead of growing in size.

Another indicator of repeasantization is the participating farmers' establishment of a community-controlled economy. This includes informal trading, giving and taking based on generosity and reciprocity, money-based transactions that are based on mutual goodwill, and production for co-sufficiency. This allows farmers to distance themselves from the capitalist market economy to meet many of their immediate needs in self-controlled and culturally appropriate ways. An indicator of the strengthening of this community-controlled economy is the growth of the Ithaca Farmers' Market and the steep increase of CSA shares in the area.

"Co-production of men and living nature" (Van der Ploeg 2008) is another common theme of the peasant condition that can be observed in the Central Finger Lakes. This covers the production of consumer goods and services (for sale or self-use) and the reproduction of the farm's own resource base, including natural and social resources. Goods and services are the most visible outputs of production. They comprise the foodstuffs, inputs for other farms such as seeds and manure, and other raw materials, as well as labor and processing services that are sold and accounted for. Production for household consumption, as well as products and services for trade or gifting also fall into this category. On the other hand, the participating farmers also work to reproduce their resource base, with the ultimate goal of preserving and improving it for future use. These efforts include land improvement and soil-conserving cultivation techniques that restore soil fertility. Farms 2, 3, 4 and 5 explicitly mentioned that they use organic cultivation methods for their positive environmental, conserving effects. The gardener goes out of her way to take care of pollinators. One of the reasons why farm 6 transitioned to organic agriculture was their concern about overuse of pesticides. In terms of reproducing social resources, the farmers actively build a supportive community, both on the farm and beyond. This co-production takes place as a means of increasing autonomy in order to shape one's life according to one's values and goals.

Another point that indicates repeasantization is the focus on labor-driven intensification on four of the visited farms. Operations 1, 2, 4 and 5 mainly rely on hiring more workers or attracting more volunteers if they choose to increase their production. This intensification leads to quantitative repeasantization, because it allows more people to work and live on the land and learn the necessary skills to become farmers. While this may be regarded is a process of proletarianization instead of repeasantization (people get hired as rural laborers instead of beginning their own farms), it offers a chance for people to gain experience with farming and acquire the skills and sometimes financial capital necessary to start an own farm (see Ekers et al. 2016 for an in-depth discussion of the fine line between empowerment and exploitation in volunteer farm labor). Interviews with interns and farmworkers suggested that this is the chosen route for at least some of them. Farm 3 increased its sales with more labor, when the operator began to concentrate on direct marketing, accepting the higher workload associated with it.

These five activities, improvement of the self-controlled resource base, struggle towards more subsistence production, establishment of a community-controlled economy, co-production of human life and living nature, and labor-driven intensification show that qualitative repeasantization is a development trajectory of the six visited farms. It is important to acknowledge, that some of the tactics employed by the interviewed farmers, purchasing land and reliance on volunteer labor specifically, are just as compatible with capitalist accumulation as they are with repeasantization. Without the additional information from the interviews, that suggest that these strategies are indeed directed towards autonomy, it would be impossible to discern the farmers' ultimate goal. Further, these strategies suggest that within a capitalist society, some people resort to tactics that are compatible with the hegemonic economic system in order to escape it.[5] This approach highlights the contradictions that the tension between internal and external conditions engenders.

Quantitative assessments of repeasantization require extensive data collection that go beyond the U.S. Census of Agriculture. Repeasantization largely takes place in the margins of agriculture: people that start a homestead, a backyard garden, or a small farming operation on someone else's farm, et cetera. These operations often fall below the farm threshold set by the USDA and are therefore not counted in the census. Yet, they can be both starting points for new farms with a focus on peasant agriculture as well as manifestations of peasant farming in their own right.

Regardless, between 2007 and 2012, the number of farms in Tompkins, Schuyler and Seneca Counties has increased, particularly small farms with less than 4 ha. The data does not imply what kind of farming (peasant, entrepreneurial, corporate, or a combination) these new farms practice or aspire towards. However, this trend suggests that livelihoods on the land are having a comeback, after years of rural depopulation. It should be noted that repeasantization is not the only agricultural development in the area. Many farms continue to industrialize, while others discontinue operation. However, this trajectory is met with a counter-development of repeasantization that is gaining momentum.

Conclusion and discussion

This research demonstrated that when understanding farms as operations that are driven (if not torn apart) by competing conditions, the theory of peasant farming becomes applicable in describing the counter-hegemonic pole of this dialectic. On the farms visited for this case study, this pole presents itself as pockets of peasantness. But what inferences can we draw from looking at six farms in a single locality about peasant farming in the United States? Although not generalizable, the challenges of these farmers speak to a broader theme: Farmers may be oriented towards subsistence, or peasantness, at the same time that their situatedness within a capitalist market society (Polanyi 1944) forces them to engage in forms of agriculture that run counter to that goal. Below the surface of accounted commodity production, there can be a range of practices and economic activities that are directed at subsistence. By setting up a dichotomy between capitalistically oriented agricultural operations and subsistence-oriented peasant farms, we miss out on the large intersection between the two, and the struggle between intrinsic and extrinsic conditions that these poles imply, often on the very same farm. In that regard, the dualism in US agriculture runs deeper than commonly assumed. The juxtaposition of a small-farming, peasant(-like) ethic and growth-oriented, capitalist agriculture does not only create a dualism between small and large farms (Friedmann 1978; Bonanno 1987; Lyson et al. 2008), but also creates friction within single operations. Thus, this study provides evidence that backs Van der Ploeg's (2008) framework of peasant-farming as a condition, or *constellation*, rather than a distinct production system.

This study highlighted the many ways in which some farmers in the Central Finger Lakes collaborate. Their cooperative arrangements do not only serve economic purposes, but also facilitate meaningful interpersonal relationships and community. This adds further evidence to Emery et al. (2017) observation that cooperation in agriculture is central to farmers' identities and well-being, and create opportunities to act collectively for transformative change. Thus, cooperation is instrumental in the production of subsistence, of which these "more-than-economic dimensions of cooperation[6]" (Emery et al. 2017, p. 1) are elements.

By highlighting independence, self-sufficiency, and social and environmental responsibility as main motivations of the interviewed farmers, this study provides new empirical

5 Bernstein (2010) argues that this is precisely why the peasantry ceased to exist. There is just no escape from engaging in capitalistic behavior. Although this observation is well taken, it does not distinguish between farmers who use these tactics for accumulation as an end in itself, or as a way to obtain the means for their simple reproduction.

6 Considering that identity, well-being, and capacities to act collectively are actively produced (through cooperation either based on shared labor or exchange of material goods) through labor, it would be more appropriate to talk about these values as more-than-material instead of more-than-economic.

evidence to recent studies that found independence (Emery 2015), autonomy (Stock et al. 2014; Stock and Forney 2014), and stewardship over land and community (Stock 2007) to be central elements of farmer's identities in developed countries.

The identification of instances of repeasantization in the Central Finger Lakes is consistent with Nelson and Stock's (2018) findings on farming in Kansas. They demonstrate that entrepreneurial farms (which would encompass the farms interviewed for this research with the exception of the garden) already employ some markers of peasantness and, thus, offer the greatest potential for qualitative repeasantization in the United States. This research from New York adds that a peasant culture on existing farms can also trigger quantitative repeasantization by inspiring the start of new, peasant-like farms. The increase of small farms in the Central Finger Lakes suggests that this might already be occurring.

This research provides some evidence that the peasant way constitutes a contemporary alternative to the hegemonic capitalist economy. Although sharing many of the progressive goals of what Stock et al. (2015) describe as "Food Utopias", the peasant conditions is decidedly non-utopian. Envisioned or prefigurative, utopias are created to bring about change to the future and present (Wright 2010). The peasant condition, however, has never been "created" but evolved over centuries. Sometimes, perhaps, it is rediscovered. But it was never gone. Neither does the peasant condition imply revolutionary intent. It is a traditional way of living that only turned into a radical alternative as it has been marginalized from the societal mainstream. From the historical perspective of the peasantry, capitalism is the radical vision, their dystopia. Pockets of peasantness allow us to study empirically the ways agrarian communities are successful, or struggle, to resist the hegemonic culture of capitalism.

The existence of a subsistence orientation among the interviewed farmers suggests that the theories of peasant studies may also be applicable to at least some sectors of US agriculture. Although the findings of this study are not generalizable to the entire region, let alone the whole country, this research proposes a way of looking at farming that may yield new insights about the substantive economy in farming communities in the US. The observation that single farms may divide their labor between peasant-like and capitalistic activities challenges the common dichotomy between large-scale, industrial, profit-driven agriculture and (allegedly) more community-embedded small-scale agriculture. Could it be, that the friction between peasant and capitalist farming does not run between farms, but within them? The idea that both large-scale and small-scale farmers may struggle to gain autonomy from corporate pressures on their operations creates opportunities for solidarity between these two sectors. That is not to say that there are no important frictions between these two classes of farms, but they do seem to share this common struggle. Future research should focus more strongly on how different constellations (e.g. corporate, entrepreneurial, and peasant agriculture according to Van der Ploeg's (2008) framework) create tension and opportunities within single farms. Especially studies that treat farming operations as their unit of analysis need to justify this analytical decision and explain how their findings may cover up within-farm dynamics. In addition, future research should investigate how different conditions, such as location, scale, management decisions et cetera, shift the equilibrium on a farm between corporate and peasant-like farming, and how these variables interact. Ethnographies that focus on specific, place-based subsistence cultures in farming communities could more strongly integrate US agriculture into the larger body of peasant studies. In addition, comparative case studies across different sites could help to derive a US-specific theory of peasant farming vis-à-vis existing models of capitalist agriculture in the United States. In order to facilitate a better understanding of peasant farming in the United States on a state or national level, it would be helpful if the US Census of Agriculture would include more questions about subsistence activities on the farms. At this point, researchers interested in peasant-like forms of farming rely on their own data collection, which impedes comparability across cases.

Taking seriously the existence of pockets of peasantness with capitalist economies implies an ontological shift (Gibson-Graham 1996) away from a monolithic understanding of (capitalist) agriculture towards diverse agro-economies (c.f. Gibson-Graham 2008). Acknowledging the existence of non-capitalist economies such as pockets of peasantness in Upstate New York (and elsewhere) means to engage in performative practice (ibid.) that creates space for alternative thinking about and acting within agriculture.

Acknowledgements This research has been made possible by a travel grant provided by University of Natural Resource and Life Sciences, Vienna (BOKU). During the fieldwork, the Department of Development Sociology, Cornell University hosted me as a visiting scholar. I thank Veronika Bennholdt-Thomsen and Rachel Bezner Kerr for their supervision and guidance. I would also like to thank Julie Dawson, Alicia Swords, and four anonymous reviewers for feedback. In particular, I would like to thank the farmers in this study for taking the time to participate and continuing their inspiring work.

References

Agar, M.H. 1996. *The professional stranger: An informal introduction to ethnography*, 2nd ed. San Diego, CA: Academic Press.

Babbie, E.R. 2016. *The practice of social research*, 14th ed. Boston, MA: Cengage Learning.

Bennhold-Thomsen, V. 2006. Subsistenzwirtschaft, Globalwirtschaft, Regionalwirtschaft. In *Lebensweltökonomie in Zeiten wirtschaftlicher Globalisierung*, ed. M.A. Jochimsen and U. Knobloch, 65–88. Bielefeld, Germany: Kleine Verlag.

Bennholdt-Thomsen, V., and M. Mies. 1999. *The subsistence perspective. Beyond the globalized economy*. New York, NY: Zed Books.
Bernstein, H. 2010. *Class dynamics of agrarian change*. Sterling, VA: Kumarian Press.
Bonanno, A. 1987. *Small farms: Persistence with legitimation*. Boulder: Westview Press.
Boyle, M. 2012. *The moneyless manifesto*. East Meon, Hampshire, UK: Permanent Publications.
Chayanov, A.V. 1966. *The theory of peasant economy*. Homewood, IL: American Economic Association.
Creswell, J.W. 2013. *Qualitative inquiry and research design: Choosing among five approaches*, 3rd ed. Los Angeles: SAGE Publications.
Edelman, M. 2013. *What is a peasant? What are peasantries? A briefing paper on issues of definition*. Prepared for the first session of the Intergovernmental Working Group on a United Nations Declaration on the Rights of Peasants and Other People Working in Rural Areas, Geneva, 15–19 July 2013.
Ekers, M., C.Z. Levkoe, S. Walker, and B. Dale. 2016. Will work for food: Agricultural interns, apprentices, volunteers, and the agrarian question. *Agriculture and Human Values* 33 (3): 705–720.
Emery, S.B. 2015. Independence and individualism: Conflated values in farmer cooperation? *Agriculture and Human Values* 32 (1): 47–61.
Emery, S.B., J. Forney, and S. Wynne-Jones. 2017. The more-than-economic dimensions of cooperation in food production. *Journal of Rural Studies* 53: 229–235.
Friedmann, H. 1978. World market state and family farms: Social bases of household production in the era of wage labor. *Comparative Studies in Sociology and History* 20 (4): 544–586.
Gibson-Graham, J.K. 1996. *The end of capitalism (as we knew it): A feminist critique of political economy*. Cambridge, MA: Wiley-Blackwell.
Gibson-Graham, J.K. 2008. Diverse economies: Performative practices for `other worlds. *Progress in Human Geography* 32 (5): 613–632.
Gray, J. 1998. Family farms in the Scottish borders: A practical definition by hill sheep farmers. *Journal of Rural Studies* 14 (3): 341–356.
Halperin, R. 1977. Introduction: The substantive economy in peasant societies. In *Peasant livelihoods*, ed. R. Halperin and J. Dow, 2–16. New York, NY: St. Martin's Press.
Kautsky, K. 1988. *The Agrarian Question. (trans: P. Burgess)*. Winchester, MA: Swan Publications.
Lyson, T., G. Stevenson, and R. Welsh. 2008. *Food and the mid-level farm: Renewing an agriculture of the middle*. Cambridge, MA: MIT Press.
Mann, S.A., and J.M. Dickinson. 1978. Obstacles to the development of a capitalist agriculture. *The Journal of Peasant Studies* 5 (4): 466–481.
Marx, K. 1977. *Capital. critique of political economy. (trans: B. Fowkes)*, vol. 1–2. New York, NY: Vintage.
Nelson, J., and P. Stock. 2018. Repeasantisation in the United States. *Sociologia Ruralis* 58 (1): 83–103.
Patton, M.Q. 2002. *Qualitative evaluation and research methods*, 3rd ed. London, UK: Sage.
Polanyi, K. 1944. *The great transformation*. New York: Rinehart.
Polanyi, K. 1957. The economy as instituted process. In *Trade and market in the early empires Economies in history and theory*, ed. K. Polanyi, C. Arensberg, and H. Pearson, 243–270. Chicago, IL: Henry Regnery Company.
Reinhardt, N. 1988. *Our daily bread: the peasant question and family farming in the Colombian Andes*. Oakland, CA: University of California Press.
Reinhardt, N., and P. Barlett. 1989. The persistence of family farms in United States agriculture. *Sociologia Ruralis* 29 (3–4): 203–225.
Rostow, W.W. 1960. *The stages of economic growth. A non-Communist manifesto*. Cambridge, UK: University Press.
Scott, J.C. 1985. *Weapons of the weak: Everyday forms of peasant resistance*. New Haven: Yale University Press.
Stock, P.V. 2007. 'Good farmers' as reflexive producers: An examination of family organic farmers in the US midwest. *Sociologia Ruralis* 47 (2): 83–102.
Stock, P.V., and J. Forney. 2014. Farmer autonomy and the farming self. *Journal of Rural Studies* 36: 160–171.
Stock, P.V., J. Forney, S.B. Emery, and H. Wittman. 2014. Neoliberal natures on the farm: Farmer autonomy and cooperation in comparative perspective. *Journal of Rural Studies* 36: 411–422.
Stock, P.V., M.S. Carolan, and C.J. Rosin. 2015. *Food utopias: Reimagining citizenship, ethics and community*. Abingdon, UK and New York, NY: Routledge.
Tisdell, C. 1991. *Economics of environmental conservation*. Amsterdam, NL: Elsevier Science Publishers, Amsterdam.
U.S. Census Bureau 1945. *U.S. Census of Agriculture. New York State and County Data*. Suitland, MD: United States Bureau of the Census.
USDA, ed. 2007. *U.S. Census of Agriculture. New York State and County Data*. Washington, D.C.: National Agricultural Statistics Service, United States Department of Agriculture.
USDA, ed. 2012. *U.S. Census of Agriculture. New York State and County Data*. Washington, D.C.: National Agricultural Statistics Service, United States Department of Agriculture.
Van der Ploeg, J.D. 2008. *The new peasantries. Struggles for autonomy and sustainability in an era of empire and globalization*. London, UK: Earthscan.
Wilson, C., and C. Tisdell. 2001. Why farmers continue to use pesticides despite environmental, health and sustainability costs. *Ecological Economics* 39 (3): 449–462.
Wright, E.O. 2010. *Envisioning real utopias*. London, UK: Verso.
Yin, R.K. 2018. *Case study research and applications: Design and methods*, 6th ed. Thousand Oaks, California: SAGE Publications Inc.

Johann Strube is a Ph.D. Candidate in Rural Sociology at The Pennsylvania State University. He is interested in questions around subsistence, food sovereignty, decolonization, and Indigenous ways of knowing within Europe and North America.

Agriculture and Human Values (2019) 36:849–865
https://doi.org/10.1007/s10460-019-09949-8

SYMPOSIUM/SPECIAL ISSUE

Action research on organizational change with the Food Bank of the Southern Tier: a regional food bank's efforts to move beyond charity

Alicia Swords[1]

Accepted: 21 May 2019 / Published online: 13 June 2019

Abstract
This paper reports on an action research project about organizational change by a regional food bank in New York State's southern tier. While the project team initially included a sociologist, food bank leadership and staff, it expanded to involve participants in food access programs and area college students. This paper combines findings from qualitative research about the food bank with findings generated through a collaborative inquiry about a ten-year process of organizational change. We ask how a regional food bank can change its approach to address root causes of hunger. Acknowledging that narrow, pragmatic definitions of hunger promote charitable responses, our collaboration is grounded in structural understandings of poverty that refuse to blame the poor or treat poverty as an accident. Decades-long economic restructuring, deindustrialization and a rise in the service economy have resulted in growing inequality and long-term demand for "emergency" food in New York State. We outline critiques by scholars and practitioners of the emergency food regime. Description and analysis of the organizational change efforts of the Food Bank of the Southern Tier combine discourse analysis, collaborative inquiry, interviews, and participant observation. Discourse analysis of the agency's strategic plans documents changes in aspirations, exposure to new epistemic communities and repertoires of actions. Interviews with participants evidence impacts of the organization's advocacy and education programs on people with lived experience in poverty. Through a participatory process, we developed a collaborative chronology of phases of organizational change. Collaborative analysis of organizational changes demonstrates new definitions of the problem, a shift in service focus, changing outcomes and increased funding for advocacy. While recognizing substantial constraints, this project contributes to evidence that food banks may shift their discourse and practices beyond charity.

Keywords Food banks · Food justice · Hunger · Organizational change · Action research · Food regime · Emergency food · Dispossession

Abbreviations

AR	Action research
FBST	Food bank of the Southern Tier
SB	Speaker's Bureau, a project of the food bank of the Southern Tier

Introduction

The emergency food system, which includes food banks, food pantries, meal programs, food rescue programs, and anti-hunger organizations, concerns critics because it addresses the problem of hunger with charity rather than rights (Poppendieck 2014). Despite ongoing economic crises, budget cuts, stagnating wages, decline in union rates, and the rise of service jobs, analysts are concerned that anti-hunger organizations mainly offer charity food and public benefit programs rather than addressing poverty, which is the root cause of hunger and food insecurity (Fisher 2017).

At the same time, food banks are at the front lines of meeting material needs for a wide population of those in poverty. Given their reach, they may have unique potential for building a movement to end poverty led by the poor. In

✉ Alicia Swords
aswords@ithaca.edu

1 Department of Sociology, Ithaca College, 953 Danby Road, Ithaca, NY 14850, USA

view of this potential, we offer a case study of the efforts of a regional food bank to move beyond a charity model. Acknowledging the contradictions of food banks, I used Action Research (AR) principles to engage a core team[1] of food bank leadership and staff in a collaborative inquiry.

This paper begins by reviewing theories of hunger and poverty. Social movements committed to ending poverty and hunger advance a theory of poverty as structural (Baptist and Rehmann 2011), which provides a challenge to food-banking-as-usual. We contrast theories that blame no one or blame people for their own poverty and hunger with structural theories that recognize that poverty and hunger are the product of damaged social systems. Evidence of contemporary political-economic decline in the United States and New York State provides context for our inquiry and reveals that the Empire State is a leader in inequality. Literature from scholars and practitioners on the emergency food regime highlights the contradictory roles of food banks as well as their potential.

This paper combines qualitative interviews with analysis of organizational data gathered through collaborative inquiry. Our inquiry documents and analyzes efforts by the Food Bank of the Southern Tier (FBST) to make changes in the organization's structure, staffing, budgeting, strategic plan and outcome measures.[2]

A collaborative analysis of organizational changes used categories adapted from Fisher (2017) and demonstrated changes in problem definition, who it serves and outcomes measured; increased funding for advocacy; and impacts of its advocacy and education on program participants. Interviews with participants of the food bank's leadership program for people with lived experience in poverty illustrate that FBST's advocacy and education program contributed to their self-concept, encouraged their leadership, and contributed to structural understandings of poverty. Finally, I examine possibilities and challenges for FBST and suggest future lines of research and action for transforming the roles of food banks.

[1] The core team includes the author, FBST President and CEO Natasha Thompson (subsequently referred to as the president), and staff, Randi Quackenbush and Lyndsey Lyman. They all contributed as described below, including by reviewing data and conclusions, making edits and comments. All errors, however, are mine.

[2] This report draws from various projects with human participants, each of which received individual IRB approval under the following IRB numbers: 1215-06, 0217-16, and 0516-03. Informed consent was obtained from all individual participants included in the study. Additional informed consent was obtained from all individual participants for whom identifying information is included in this article.

Literature review

Theories and definitions of hunger and poverty

Despite widespread agreement that hunger is deplorable, contemporary understandings of this social problem focus on narrow and pragmatic definitions (Fisher 2017). The USDA (2013) defines food insecurity as the "limited or uncertain availability of nutritionally adequate and safe foods or limited or uncertain ability to acquire acceptable foods in socially acceptable ways." As Fisher (2017) and Poppendieck (1999) argue, popular definitions of hunger as lack of access to food promote voluntary, charitable responses like food drives and soup kitchens, rather than social changes that address the root causes of hunger.

Similarly, official poverty measures[3] define who is experiencing poverty but do not identify causes of poverty. Taken at face value and without relational understandings, such measures are used to imply that individuals can "get out of poverty," reinforcing common cultural assumptions that suggest the poor should pull themselves up by their bootstraps. Popular and predominant explanations for poverty blame the behavior of the poor or treat poverty as a temporary accident (Goldsmith and Blakely 2010). Such assertions frequently reinforce racist or classist assumptions. Cultural references reinforce these theories, such as the common phrase that someone is "down on their luck" or the caricature of the Welfare Queen, a black single mother who gets rich from her welfare checks (Zucchino 1997).

In contrast, sociological and historical analysis suggest that poverty, and therefore hunger, occurs as a result of structures built into the economy and society (Goldsmith and Blakely 2010; Baptist and Rehmann 2011). Under capitalism, poverty is a necessary byproduct of capital accumulation; the creation of wealth for a few requires the impoverishment of many. In Capital Volume 1, Marx defines primitive accumulation as the process of evictions, colonialism and theft that allowed the emerging capitalist class to accumulate wealth by stealing the land and the labor of others (1867). For Harvey (2003) accumulation by dispossession emphasizes the ongoing nature of dispossession, via privatization of public land and public services, financialization and manipulation of food prices, and structural adjustment. A structural approach argues that poverty is a product of historical, social relations, that policies contribute and

[3] The official poverty measure is three times the inflation-adjusted cost of a minimum food diet, based on average family expenses in 1963 (U.S. Census Bureau 2017). The supplemental poverty measure (SPM) includes cash resources and noncash benefits from government programs (U.S. Census Bureau 2017). The Census also calculates deep poverty (Shaefer et al. 2012).

exacerbate it, and that dynamics inherent in the economic system impoverish and dispossess people. Further, structural explanations clarify that individuals who are poor are not to blame for their own poverty. For this reason, poor people's organizations teach structural theories of poverty as part of consciousness raising (Baptist and Rehmann 2011). They suggest that poverty is not individual, but an ill of the whole society.

As this AR project developed, I introduced structural explanations for poverty and hunger to FBST leaders and staff and in workshops as part of a leadership training for participants in food access programs, as described below. Our research teams, including college students and food access program participants, used a structural framework (Goldsmith and Blakely 2010; Baptist and Rehmann 2011) for the analysis that follows.

Economic crisis, growing inequality, and poverty: impacts in New York State

Economic restructuring, growing inequality and poverty have their origins in long-term crises in agriculture and manufacturing. The "emergency food system" emerged in the 1980s with recession, economic collapses in industrial cities, and Reagan-era cuts in government entitlement programs (Fisher 2017; Poppendieck 1999). Economic restructuring and austerity policies shrank middle-income strata and swelled the ranks of the poor, exacerbating inequalities that have made the "emergency" a long-term, structural reality.

Economic restructuring occurred as wages and productivity in the United States decoupled in the 1970s; real wages largely stagnated while productivity continued to rise (Mishel 2012). Corporate leaders attempted to resolve periodic crises of overproduction via wage cuts, deunionization, subcontracting, flexibilization and the increasing use of immigrant and women's labor (Robinson 2008).

A significant contribution in New York's Southern Tier during this time period was made by Janet Fitchen (1981, 1991). A "social scientist-activist," she referred to the region as the northern tip of Appalachia and defined rural New York State based on its land use in agriculture, forests and underdeveloped countryside, and by social identity or state of mind. Her work contributed an understanding of social consequences of farm crisis and manufacturing job loss on rural poor people. In 1985–1990, as she conducted fieldwork, she warned about the impact of cuts in federal services on rural culture. She argued that loss of a manufacturing plant in a small community "packs a much stronger wallop" (p. 71) than the same closure in an urban area.

One dimension of economic restructuring was deindustrialization, a widespread trend of disinvestment in productive capacity (Bluestone and Harrison 1982). Another was the rise in service-sector jobs. Two major episodes of manufacturing job loss in the U.S. occurred, in the 1980s and 2000s (Atkins et al. 2011). From 2000 to 2014, the U.S. lost over 5 million manufacturing jobs (Scott 2015). New York State, by 2000, lost over 60% of its manufacturing jobs since its peak in the mid-1940s (DiNapoli 2010). From 2000 to 2008, upstate New York's manufacturing sector lost 105,000 jobs, declining by 20% from 2000 to 2004, and then by 8% from 2005 to 2008 (DiNapoli 2010). In 2008 and 2009, New York was hit by the Great Recession and lost more than 300,000 jobs (DiNapoli 2013). After the recession, from 2007 to 2012, the only economic sectors that grew were service sectors, with the most growth in leisure and hospitality and education and health (DiNapoli 2013). Yet these new jobs did not provide the wages manufacturing jobs once did.[4]

Although New York State's 2016 gross state product was nearly $1.49 trillion (U.S. Bureau of Economic Analysis 2017), enough to put it among the highest ranking 13 *countries* in the world, it is the most unequal state in terms of wealth distribution. Since 1979, the average incomes of the top 1% have grown by 272% in inflation-adjusted terms in New York, while the average incomes of the bottom 99% rose a meager 5.4%" (Sommeiller et al. 2016). According to the Economic Policy Institute, New York State has the highest rate of income inequality in the country (Sommeiller et al. 2016).

In 2016, the official poverty rate in New York State was 15.7%. The United Way's ALICE (Asset Limited Income Constrained Employed) Threshold depicts the gap between the poverty line and regional costs of housing, childcare, food, transportation, health care and taxes.

In New York State, there are 2.1 million ALICE households, while another 1.1 million households live below the poverty level. In total, 44% of the state's households earn below the ALICE Threshold, meaning that they cannot afford housing, childcare, food, transportation, healthcare and taxes (United Way 2016).

The reduction in steady, well-paying jobs has spurred reliance on food distribution systems once thought to be temporary and for "emergencies" only. Federal emergency food assistance programs, which were born around the time of the Great Depression, have been supplemented by privately funded food banks, food pantries and soup kitchens (Poppendieck 1999). More and more individuals and families rely on food assistance programs. Feeding America, a U.S. network of 200 member food banks, reported distributing 4 billion meals in 2016, nearly tripling its annual food distribution since 2008 (Feeding America 2008, 2016).

[4] From 2009 to 2014, wage growth in leisure and hospitality was moderately above the inflation rate education and health was only moderately above (DiNapoli 2015).

In a 2012 study of emergency food programs in New York State, more than 80% of programs reported an increase in the number of clients over the prior year. Half reported an increase in the number of working people and seniors, and 75% reported an increase in children (St. Clair and Dunlea 2012). In their six-county service area, according to Feeding America data, FBST provided more than 8.3 million meals, but the estimated demand was for 5 million more (Gundersen et al. 2016). Food insecurity is a regular concern of 12.5% of New Yorkers and 13% of people nationally (Coleman-Jensen 2017).

The emergency food regime: critiques and contradictions

Scholars McMichael and Friedmann defined food regimes as constellations of class relations, geographical specialization and inter-state power that connect international relations of food production and consumption with periods of capitalist accumulation (Friedmann 1993; Friedmann and McMichael 1989, p. 95; Friedmann 2009, p. 335). Food regimes include actors such as states, corporations, scientists, social movements, consumers and others. They adopted the term regime from Gramsci (1971) to emphasize that these are configurations of social forces that achieve temporary relative stability. Today, according to McMichael (2005, 2009), the industrial food regime is giving rise to a corporate regime based on corporate control and financialization. The framework of food regimes invites investigation of what might shift the social constellations of the contemporary food regime. Here we focus on the *emergency* food regime, with an appreciation that these social relations may be temporary.

Criticism of the charity model in food banks is well established. Poppendieck's (1999) *Sweet charity? Emergency food and the end of entitlement,* defined the field. She outlines the seven deadly "ins" of food banks (a play on words referring to the Catholic concept of seven deadly sins). Emergency food institutions, according to her research, offer food that is (1) insufficient in quantity, (2) inappropriate or mismatched with consumer choices, (3) nutritionally inadequate. Their providers struggle with (4) instability, (5) inaccessibility, (6) inefficiency, and (7) reproduce indignity. Most of all, emergency food programs become part of a vicious cycle. They were created to compensate for the limitations of public entitlements, but actually further undermine them by providing a "moral safety valve" to the public (Poppendieck 1999, p. 301). For example, in eliciting donations, food banks often argue in favor of their efficiency over that of public programs. The very act of offering food distribution as an "anti-hunger action" precludes approaches that better address the root causes of hunger and poverty.

McIntyre et al. (2016) analyzed publications that built on and extended Poppendieck's seven deadly [s]ins. Their analysis contributes two constructs to distinguish between contemporary critiques of food banks: operational challenges and perpetuating inequalities. They also show how the literature has extended Poppendieck's critique with these five general arguments: (1) Food banks are ineffective in addressing individual and household food insecurity. (2) They create or reproduce inequality among donors, volunteers and recipients. (3) They promote institutionalization of food provision via bureaucratic procedures. (4) They invalidate entitlements. (5) They make hunger invisible by creating the impression that the problem has been resolved.

Most challenging for food banks is the critique that they perpetuate inequalities. According to such critics, a narrow focus on hunger as a social problem promotes pragmatism rather than actions to address root causes. The proliferation of charity relieves pressure and legitimates personal actions as a response to economic dislocation (Poppendieck 1999). Further, Fisher (2017) argues that the narrow, pragmatic definition of hunger as the lack of access to adequate food ignores the relationship between hunger and poverty and "acts as a barrier to its own elimination" (p. 20). Poppendieck (1999) argues that a number of interest groups benefit from the emergency food system, including the U.S. Department of Agriculture, businesses, churches, environmentalists, as well as groups, organizations and institutions that benefit "from the halo effect of 'feeding the hungry.' If we didn't have hunger, we'd have to invent it." (Poppendieck 1999, p. 293). Hunger alleviation allows people to satisfy religious and cultural obligations and offers a simple way to alleviate guilt and prevent food waste without confronting larger systems of inequality.

Fisher (2017) critiques the alignment of anti-hunger organizations with large corporations in an "unholy alliance" he calls the anti-hunger industrial complex. Food banks channel charitable donations via food pantries to distribute donated food as well as surpluses purchased from food processors and retail chains. Because corporations receive tax write-offs in exchange for surplus foods and anti-hunger donations, food banks provide positive public relations for corporations. By "subsidizing mainly processed, boxed, and canned commodity producers for their mistakes (mainly over-supply or mislabeled goods), the emergency food system benefits big industry and the poverty-industrial complex of nonprofits" (Vitiello et al. 2015, p. 420). The board of directors of Feeding America, a national network of food banks, includes representatives from major food, agriculture and finance industries, including Morgan Stanley, Walmart, The Kroger Company, and ConAgra Foods (Feeding America 2017). With leaders and funders beholden to corporations, anti-hunger organizations find they must avoid rocking the boat, so they are further constrained in addressing root causes of hunger. Winne (2008) finds that generally food banks do not lead critical public discussion around hunger and poverty "because influential people don't attain exalted positions within a community's hierarchy by asking hard, controversial questions or becoming agitators" (p. 76). As well, he finds that boards and officers of national food banks usually include CEOs or high-ranking officers of major corporations. "Rarely do you find a person of color or someone who has received assistance from a food bank among them" (Winne 2008, p. 76).

Despite the depth of their critiques, Poppendieck (1999), (2014), Winne (2008) and Fisher (2017) envision roles for emergency food providers in transforming the emergency food regime. Poppendeick (1999) calls for food banks to participate in a movement. They should not focus uniquely on hunger, she explains, but rather on poverty. He asks what might happen if food banks "put all the effort soliciting and distributing wasted food into ending hunger and poverty?" (p. 77).

Recent research examines food bank efforts to get closer to the roots of hunger. Galinson (2018) examines two case studies of California food banks that added advocacy programs, suggesting that such programs can reduce hunger and address its root causes in poverty. She argues, that rather than working themselves out of a job, as they aspire to, "they've worked themselves into a new and necessary job: pushing food accountability back on the government, and in the most progressive cases, using advocacy to enrich their communities by ensuring those affected have a voice in creating solutions" (p. 97). Two interconnected action research projects by Dodd and Nelson (2018) work with service providers and service users to shift discourse and practice about emergency food provision. Our collaborative inquiry aims to contribute to such studies of food bank potential by engaging leaders and staff of a regional food bank in examining organizational change.

Methods: action research for organizational change

Action Research (AR)[5] is a practice that combines research, action and education to addresses real-life problems in a holistic manner (Greenwood and Levin 2007, p. 63). It involves participants, or co-investigators, with diverse experiences and capacities in seeking solutions to problems which are tested in practice (Greenwood and Levin 2007, p. 111). A core commitment of AR is to generate social change to ease oppressive conditions with participation by those directly affected. Ideally, participants are involved in all phases of knowledge production, from problem identification, research design and methods, dialogue, data gathering and analysis, and utilization (Park 1993). Embedded within broader social change processes, AR can support people facing oppression to develop critical consciousness, identify obstacles to change, and challenge oppressive structures (Greenwood and Levin 2007).

In North America,[6] AR related to poverty and hunger has been used to identify issues that affect participants and generate social action or policy change. While techniques and technologies differ, AR combines knowledge from researchers and participants and encourage individuals to participate in efforts toward raising awareness about the issues facing them (Collins 2005; Wang 1999; Valera et al. 2009; Knowles et al. 2015). Their projects frequently involve connections between communities and universities (Valera et al. 2009; Collins 2005), and some, like ours, build multi-year community-university partnerships (Dodd and Nelson 2018).

The tradition of using AR to facilitate organizational change draws on social psychologist's Kurt Lewin's efforts to involve workers in improving their conditions via interventions to undo old structures, change structures, and replace structures (Greenwood and Levin 2007). McNiff and Whitehead (2000) suggest that organizations can be learning organizations by developing "generative transformational theories", which includes "studying their own changing work and telling the stories of their learning processes as they tried to make a difference" (p. 57).

This project describes organizational change by a food bank, focusing on food bank practices. We ask, as McNiff and Whitehead (2000) suggest, how a food bank can understand and develop its work. We combine knowledge from a researcher, food bank staff and participants with lived experience of poverty to investigate the organizational changes made by a food bank. Similar to Dodd and Nelson (2018), we examine changing discourse and practices and work with both service users and service providers.

In 2011, I met the president of the Food Bank of the Southern Tier. Prompted initially by our shared commitment to the goal of ending poverty, we began discussions about a possible collaboration. As we reflected on an initial service-learning project and a visit with the Poverty Initiative, a non-profit organization focused on building a social movement to end poverty, we agreed to use an Action Research approach toward a shared goal of moving FBST beyond a charity model. Over the last 7 years, we have built and expanded the collaboration through civic engagement and service-learning projects with college, faculty and student

[5] Action Research, Participatory Research, Participatory Action Research, and Community-Based Research are a cluster of practices that all share common principles. They cross disciplines and address a range of problems, but all combine education, research and action (Hall 1979b in Park et al. 1993).

[6] Many methods to address on poverty and hunger were developed in international development contexts.

involvement in FBST initiatives, as well as participation by the agency in the national movement to end poverty.

Our efforts have gradually involved more stakeholders. The FBST president invited staff from the FBST to participate, as well as hiring new staff for key positions, and supported development of the Speaker's Bureau, an education and advocacy program to support and develop the leadership of food pantry users and people with lived experience in poverty. As well, I developed two sociology classes which participated in the inquiry.[7] Today, the collaboration involves the food bank president, three staff members, ten members of the Speaker's Bureau, and myself, a sociology professor. Forty-eight college students have been involved (four classes of twelve students each).

AR processes often involve iterative cycles of inquiry, action and reflection. In the best cases, they can be considered a spiral, with iterative inquiry that builds knowledge through the process (Greenwood and Levin 2007, p. 112). Our process has developed through a series of collaborative inquiries (Kelly et al. 2001), with each phase of inquiry involving new theories, knowledge, action, and reflection, building upon prior phases. Each phase of the inquiry has focused on specific goals. An initial goal was to develop a hunger education curriculum for FBST. Next, we worked together to learn more about the experiences of people who use FBST's pantries. Later, we have worked to build involvement by people with lived experience in poverty and those most affected by hunger. By the second phase, we knew our project required a clearer understanding of root causes of hunger and poverty. By 2015, it was clear that our collaboration would involve collaborative institutional analysis to analyze changes in the food bank's practices and to advance such changes.

Action Researchers propose that those affected by research should participate in the process of the research. Existing typologies of participation (Arnstein 1969; Pretty 1995; White 1996) carry implicit normative assumptions that suggest a progression from superficial or manipulative to "genuine" forms (Cornwall 2008). As Cornwall argues, participation shapes and is and shaped by power relations, and that meanings of participation include both intentions and actions of those who initiate participation as well as interests and actions of those who participate. Therefore, rather than aiming for "full participation" or assuming deep and wide participation, we follow Cornwall's suggestion to aim for "optimal participation", which means finding an appropriate balance between between depth and inclusion for the purpose and context.

This paper combines data gathered through qualitative sociological methods as well as through collaborative inquiry. With the guidance of the core team, I collected organizational documents, including job descriptions, strategic plans, newsletters, grant proposals, and other organizational texts. With consent, I kept personal communications and meeting notes from meetings with the core action research team. A graduate research assistant and I conducted interviews with six participants from the Speakers Bureau and attended the SB public graduation ceremony. Students in 2016 and 2017 conducted participant observation at eight pantries and interviewed ten pantry volunteers and coordinators. To document the inquiry and organizational learning, the core team developed a chronological chart. Our collective analysis of this data used theories of poverty and Fisher's (2017) categories about food bank changes to assess the FBST's changes.[8] As well, my graduate assistant and I used narrative analysis (Esterberg 2002), guided by feminist interview strategies (Devault 1990) to analyze interview transcripts and notes, identifying common themes, concepts, structures, identities, conflicts and meanings of their stories, which we then checked with the core team.

In this collaboration, I have aspired to serve as "friendly outsider", combining critique and support, showing opportunities for change, and making evident tacit knowledge (Greenwood and Levin 2007, p. 125). With FBST, I have served as collaborator, volunteer consultant, outside critic, and academic partner for service-learning.[9] My role has always been informal, uncompensated, and evolving. I have shared theoretical models that see the root causes of hunger as structural poverty and have encouraged FBST to learn from organizations that demonstrate a structural analysis, including a national network of poor people's organizations. Five of my sociology courses have engaged in service-learning projects with FBST, in which students conducted research about food pantry volunteers' attitudes and behaviors and developed sociological evidence regarding issues

[7] These were senior seminars in sociology at Ithaca College. I taught Community Organizing in 2012 and 2015, and Inquiry and Action for Social Change in the spring semesters of 2016 to 2018.

[8] While conventional social research seeks objectivity, action researchers follow the hermeneutical position that reality is subjective and social science aims to interpret reality (Greenwood and Levin 2007 p. 68). There is no set of methodological rules that can substitute for testing knowledge generated in practice. Therefore, our process was an interpretive one, aiming to achieve credibility and workability (Greenwood and Levin 2007, p. 81).

[9] In Summer 2012, I collaborated with FBST's curriculum committee to design their Hunger Education curriculum. I consulted with FBST staff to plan community focus groups with clients in Summer 2015, consulted to develop the Speaker's Bureau training curriculum (from 2016 to present) and to design a management team training on root causes of poverty (Spring 2018). I conducted workshops on root causes, realities and theories of poverty for FBST staff and for SB trainings in Spring 2016 and 2017.

selected from pantry users' stories.[10] Through this collaboration, I have learned a great deal from FBST leaders, staff and participants.

Findings

Discursive change in FBST's strategic plans

One way to assess change is to examine the organization's explicit discourse, or language used to describe its work, mission and goals. Four strategic plans were available, for 2005, 2007, 2010–2012, and 2017.[11] Based on categories for analysis that we discussed and agreed upon in our collaborative inquiry, I analyzed the discourse of these plans to assess the organization's projected role and theory of change, the organization's proposed goals, and outcomes or measures of success. I then reviewed my observations with the core team.

In the early plans, despite its non-profit status FBST projects itself as a business focused on food-distribution. The 2005 plan included categories adopted from the business sector including Warehouse Operations, Business and Support Services, and Development and Public Relations, in addition to the social service category of Agency and Program Services. The 2007 plan projected a new focus on "nutritional quality as much as on pounds distributed", while continuing to emphasize "fundamental business practices, technology and information management, corporate governance and risk management."

The 2010–2012 plan was titled a "business plan." Its language continues to reflect a business model, including terms such as "facilities transition" "financial growth," "brand awareness," "sourcing-distribution pipeline," and "operational excellence." Their mission as "the leading anti-hunger organization in the region," begins with the aspiration that "no one should go hungry." Within the context of language of business excellence, this hardly raises an eyebrow, but should be read as a commitment to distribute food effectively.

The 2017 plan, however, includes different language, reflecting a new ethic as well as FBST's involvement in a new epistemic community, or a new network of knowledge-based experts (Haas 1992). The plan is organized around three axes, Feed, Lead, and Strengthen, each focused on an aspiration, and anchored in the "two feet" of Catholic social teaching: justice and charity (United States Conference of Catholic Bishops 2018). "FEED" aspires to "address the problem of hunger today" by providing "quality, nutritious food to those in need." The second axis,"LEAD," strives to "Mobilize a movement around the issue of hunger", through education, advocacy, convening and collaboration. In particular, the education goal proposes to "develop a Common Message Framework that includes local stories, data and research" and "increase general awareness about the root causes of hunger and debunk stereotypes." The third axis is to strengthen internal capacity (i.e., staff competencies) and build capacity of member agencies to carry out the first two strategic goals, through technical support, training, incentives and funding for agencies willing to improve or expand services. The first axis maintains the language and commitment to efficient business practices. The third refers to "training" and "services", terms which are broad enough to bridge business and non-profit or social service ethics. However, the second axis adopts terms used by social movements including "mobilize", "movement", "advocacy", and "awareness." This second access demonstrates FBST's engagement with a new epistemic community, beyond anti-hunger organizations.

In combination with our participatory data, below, the language in FBST's strategic plans reveals that the organization is engaged in a new epistemic community, including a network of social movement organizations committed to ending poverty. Its 2017 strategic plan describes new analysis regarding the problem of hunger. It acknowledges that the "root causes" of hunger are in poverty and inequality and that addressing these requires a "movement." The plan also expresses a desire to add a new repertoire[12] of actions which we call a justice food bank repertoire, including educational workshops, advocacy, and mobilizing. While FBST's strategic plans demonstrate a changing vision and aspiration, their impacts must be assessed in practice. Additional data allows us to consider the practical implementation of this vision.

[10] In 2012 and 2015, Community Organizing classes (senior sociology seminars) developed and piloted hunger education workshops for high school students, educating students and themselves on the realities of poverty and hunger. Later, I developed a 400-level Action Research class called Inquiry and Action for Social Change. In Spring 2016 and 2017, students in this class interviewed volunteers about their perceptions about clients; observed social interactions and conducted preliminary political-economic research. In Spring 2018, students developed sociological briefing papers for the FBST and SB graduates to provide sociological data about issues selected from SB graduates' personal stories.

[11] Strategic plans from 2005, 2007, 2010–2012 and 2017 were provided by the current FBST president. The first two were implemented under the prior president.

[12] Social movement scholar Charles Tilly (2002) defines repertoires of contention as a set of performances by which any pair of politically constituted actors make claims on each other, which if realized, would affect their object's interests.

Advocacy, education and its impacts on participants

From 2005 to 2015, the organization more than doubled its annual revenue, from $6 million to $14 million. It increased the in-kind value of food donated to the agency and secured greater fundraising revenue. In turn, it tripled its staff, adding warehousing positions to support its increased food distribution, and fundraising and programmatic positions to support pursuit of the advocacy and educational goals and client-focused community organizing model outlined in the most recent strategic plan. The President and Chief Operating Officer explain that this financial growth is a factor that has enabled the FBST to broaden its social justice mission. However, Fisher (2017) warns that the growth mindset keeps food banks beholden to corporate donors.

Like many food banks, FBST documents increasing budget for advocacy and education, from $65,632 in 2012 to $221,789 in 2017. In other food banks, advocacy programs are often housed in "external affairs" offices, along with fundraising and external communications, but FBST's president decided that advocacy is not about making the organization look good. In 2015, FBST established a separate advocacy and education department which as grown to four full-time employees (personal communication, Chief Operating Officer, FBST 2018).

Fisher (2017) recommends that Food Banks increase their advocacy on public policy to at least 10% of organizational budgets by 2022 (p. 75). He proposes that advocacy should focus not only on charitable safety-net programs but on policies to address income inequality and poverty, including minimum wage, affordable housing, health care, labor relations, fair tax and job creation. In this spirit, the FBST has gotten involved with networks of organizations that focus on the root causes of poverty and hunger. As well, an initiative of its advocacy and education department, the Speaker's Bureau, aims to build leadership among people with lived experience in poverty.

Preliminary impacts of FBST's changes are evidenced in interviews with program participants. A graduate research assistant and I conducted interviews with six Speaker's Bureau participants and six other clients in June and July 2016, and I also attended the public Speaker's Bureau Graduation in May 2017 where they delivered public testimonies. With narrative analysis (Esterberg 2002) and feminist interview strategies (Devault 1990), we identified common themes including the impact of the Speaker's Bureau on participants' self-esteem and sense of purpose, theories of poverty, and motivation for leadership.

The opportunity to interact with others in similar situations and experience their own leadership helped to build their self-esteem, develop a sense of purpose, and challenge internalized oppression. For several who were domestic violence survivors, the most significant impact was "Realizing that we are not all alone. We don't have to be ashamed, and we don't have to stay behind closed doors." One participant exclaimed, "Thank you for giving me a purposeful life again. I am filled with hope and plan to spread this hope to my community, region, and Nation; and just maybe, to the whole World!" (Speaker's Bureau Participant, April 2017). After meeting speakers from the national Poor People's Campaign: A National Call for Moral Revival, one SB graduate explained, "I just get so doubtful of myself, but after listening to [speakers from the campaign] I feel lifted and just have a lot to say. I'm definitely not 100% confident but it sure did grow an extra 20%."

One impact of the SB was to change the explanations for poverty held by participants. Most described a prior belief that poverty and hunger were their own fault. By the end of the workshop series, participants frequently contrasted their explanations for poverty and hunger with those held by others.

> "It's easy to lump 'the poor' all together like it's one thing. A blight, a shame, a problem that doesn't go away. They'd like to sweep us under a rug, shoo us away like gnats…They think we don't want to work, pay our own way, succeed in life. They think [poverty] is a choice we made."

The term "people with lived experience" has become commonly used among FBST staff and SB participants and seems useful in helping people acknowledge their experiential knowledge and their own expertise. As well, they use the term "context expert" to recognize the importance of experiential knowledge, in contrast with "content experts," who are "professionals, service providers, and leaders with formal power who have knowledge, tools, and resources" to address issues (Attygale 2017). SB participants are encouraged to tell "their stories," during the SB workshops and are encouraged to share them with the public at the SB graduation. Their stories are frequently told as stories of struggle and survival. Some also narrate an experience of coming to awareness and joining an organization, a form of story frequently cultivated in social movements (Marshall Ganz 2011; Reinsborough and Canning 2010). The SB training has included explicit training in public speaking and telling their stories, drawing on a variety of inputs, including Toastmasters and a Story Circle led by the local organization Civic Ensemble drawing on Roadside Theater's Story Circle techniques (Roadside Theater 1999).

Participants acknowledged that their initial motivation was due to material incentives.

> "They explained the concept of a 'Speakers' Bureau' and offered me a role. I said 'yes,' but I had no idea what they were talking about. I heard 'stipend' and 'free classes' and the possibility of travel, etc. I came

primarily for the incentives at first, I must admit" (Interview, July 2016).

Gas cards and stipends clearly make ongoing participation possible. A frequently cited benefit was also the camaraderie and connection they built with the other women. One participant stated, "We have supported each other and held each other through a lot of stuff."

After participating in the SB training, nearly every person interviewed described a desire to give back to the pantries that had helped them, to their communities and to the broader society. One participant explained,

> "The advocacy I've done in the past is on a smaller scale, more of a one-on-one thing, and I love it, but to be a leader, I want to be able to say that I can... do things that are going to have a positive impact on a broader range of people" (SB Participant, July 2016).

The SB training also instilled the desire to influence elected officials. When asked what they wanted elected officials to know, one participant exclaimed, "Get off your asses, get out from behind your desks. Start really seeing the people and hearing the people and actually make some changes in the system" (SB Participant, July 2016). Participants often expressed the hope that their stories could offer necessary information and awareness for policymakers.

> "I just think they need to wake up and see reality...that the people that are using these pantries aren't leeches, parasites, or jobless bums... Like, come do my job for a day or live the way that the people that visit the pantry and I live before you try to cut funding or tell us that it's not needed." (SB Participant, July 2016).

A related theme was that the SB allowed participants to visualize their potential impact in their communities. One described the opportunity to speak on a panel about her experiences of poverty and hunger and lobby legislators during the March 2017 Feeding America's National Anti-Hunger Policy Conference in Washington, DC. She later wrote,

> "In the first week of March my daughter and I had the privilege to attend and speak at Feeding America's annual National Anti-Hunger Policy Conference in Washington, DC... We were asked to speak because we are 'persons with lived experience' who've been trained to speak, through the agency of the Food Bank of the Southern Tier, as their Speakers Bureau, to raise awareness about the issues of poverty, especially hunger. And we're advocates to end the crises as soon as possible. To do this we ask to be heard by the public, and by our elected representatives, and our stories to influence policy in organizations and Government. And we hope to save programs that are working; without which, much and grievous suffering will continue to accelerate" (Rogers 2017).

In addition, FBST has made steps to include SB graduates in leadership positions in the organization. One SB grad was invited to join FBST's board of directors. Another was hired for an Americorps-VISTA position on the staff of the organization. Both make valuable contributions to FBST's decision-making. Others have found volunteer opportunities with FBST and its network and are hopeful that their involvement will lead to paid opportunities.

A common theme was that participating in the SB workshops was invigorating at the same time as it was tiring. While presenting in a sociology class, one SB member said, "I love the Speaker's Bureau but I'm tired of telling my story." Other participants mentioned demands on their time, including work, childcare, going to food pantries, requirements by other social service agencies. It is not unusual for initiatives involving participation by service recipients to create fatigue (Cornwall 2008).

FBST is developing methods for evaluating the impact of the SB on participants and on the organization. Even so, preliminary evidence of participant narratives suggests that the advocacy and education program has contributed to participants' sense of purpose, clearer understanding of poverty as a structural issue, motivation for leadership and opportunities for involvement.

Phases of organizational change

The next section of data was compiled based on a participatory process with our core team: myself, the FBST president and key staff. We collectively analyzed the phases of inquiry which we compiled in a chart (see Table 1). For each phase, we examined who participated, the knowledge and theories that influenced our work, shared questions and goals, organizational changes and key events, methods we used to acquire new knowledge, and new questions that emerged. I developed the following narrated summary of the chart, which we circulated to ensure accuracy.

Phase 1

In 2012, the FBST President was featured on a public panel with national leaders of the movement to end poverty. She and I met and began to discuss a collaborative inquiry, acknowledging the importance of a structural analysis of poverty.

Our initial collaboration focused on collaborating to teach the public about causes of hunger. We agreed to build a team at the FBST, develop their awareness of structural causes and possibilities for ending hunger, and involve them in expanding the food bank's Hunger Education program. The

Table 1 Collaborative inquiry and phases of organizational change

When	Participants	Inputs: new knowledge and theories	Questions and methods	Organizational changes/key events	Critical new questions
Phase 1					
2012–2014	President, key staff, AS	President takes part in Pedagogy of the Poor panel with Poverty Initiative Data on poverty as structural Poverty Initiative immersion program	How to teach the public, especially area youth about causes of hunger? Interdisciplinary committee of regional experts selected by the president	Hunger education position Curriculum committee Poverty Initiative immersion program and workshop at annual agency conference Hamilton Speaker's Bureau Leader keynotes annual agency conference	Why is there hunger? What are misconceptions about hunger held by staff, volunteers, the general public? How to raise awareness of structural causes of hunger and poverty?
Phase 2					
2015	President, key staff and AS	Collective Impact training Key staff join Closing the Hunger Gap Leadership Team President and AS attend Liberation Economics training: social movement approach to ending poverty	What are clients' experiences at food pantries? Focus groups with pantry users from each county	Key staff attend Closing the Hunger Gap conference Creation of advocacy and education position Systematic listening to pantry users	What are best practices in social justice food banking? What are the daily lived experiences of people who experience poverty and hunger?
2016	President, key staff, AS, students	Narrative change; Leadership development	Develop Speakers Bureau for FBST. How can clients be trained to tell their stories of hunger and poverty? What are attitudes and behaviors of pantry volunteers and coordinators regarding poverty and pantry users? Student interviews and participant observation at pantries.	Establishment of Speaker's Bureau Speaker's Bureau workshop series and graduation, Generation 1	How does FBST practice change with awareness of pantry users' experiences? How can FBST cultivate leadership of pantry users? What educational processes are required to change pantry volunteers' attitudes and behaviors?
Phase 3					
2017	President, key staff and AS, students, SB grads (Generation 1 and 2)	Poverty as structural	Analyze attitudes and behaviors of pantry volunteers and coordinators Interviews Participant observation	New strategic plan Key staff attend Closing the Hunger Gap annual conference. Speaker's Bureau workshop series and graduation, Generation 2 FBST receives requests for how to begin Speaker's Bureau programs from 22 + food banks. Creation of community education position	Challenges of jealousy, competition, economic survival for SB participants Awareness of dominant attitudes and behaviors What processes are required to change FB structures and practices?

Table 1 (continued)

When	Participants	Inputs: new knowledge and theories	Questions and methods	Organizational changes/key events	Critical new questions
2018	President, key staff and AS, students, SB grads (Generation 1 and 2)	Trauma Informed Care training for staff Workshop series for FBST Management Team on Poverty as structural		SB grad joined FBST Board of Directors SB grad joined FBST staff as AmeriCorps VISTA FBST Management Team Workshop series	How can involvement of participants change the FBST?

organization created a hunger education position and created an interdisciplinary curriculum committee of regional experts selected by the president to develop a hunger education curriculum.

Questions arose as it became clear that definitions of hunger and theories of its causes were uneven even within the curriculum committee. Even so, FBST staff began to implement the curriculum and students piloted workshops in area schools. We focused on misconceptions about hunger held by staff, concerns about the charity model, and on the approach, including data and pedagogy, that could best raise awareness about the causes of hunger.

In January 2013, FBST engaged in its first collaboration with the movement end poverty. The president and hunger education staff worked with the Poverty Initiative of Union Theological Seminary to host a poverty immersion program, bringing students from the seminary and community leaders from partner organizations in the northeast U.S. to learn about social realities in the Southern Tier. The immersion leaders asked FBST how their organization could be part of a social movement to end poverty. The experience helped the president identify staff and volunteers who already held a structural analysis and raised a question that would carry through all the phases of research of how to best cultivate awareness of structural causes of hunger and poverty.

Phase 2

In a second phase, from 2015 to 2016, the president and the core team began to seek and incorporate new connections and information to answer the emerging questions. FBST began to dedicate staff time and resources to learn about best practices in social justice food banking. Staff members attended the Closing the Hunger Gap conference hosted by the Oregon Food Bank and learned about their Voices Project, which foregrounds pantry users' experience and opinions in developing respectful interactions at pantries. The president and staff visited The Stop Community Food Centre in Toronto, known as an example of a community center model of food bank. Staff began to reference innovative anti-hunger organizations, including The Stop, DC Central Kitchen, Collective Impact, and others. They also actively engaged with critics of the charity food banking model, notably sponsoring a community forum in March 2018 with Andy Fisher, author of *Big Hunger: The Unholy Alliance Between Corporate America and Anti-Hunger Groups* (2017).

With the support of the president, new ideas and models were shared not only within the organization but across the agencies served by the FBST. In 2015, FBST staff participated in a training about Collective Impact (CI) (Kania and Kramer 2011), a model for coordinating efforts across government, business, philanthropy, non-profit organizations and individuals, and invited a CI leader to keynote the agency conference. Key staff participated in the conferences and joined the leadership team of Closing the Hunger Gap, (https://thehungergap.org/), a network of anti-hunger organizations including food banks that challenge the corporate approach of the broader Feeding America network and are committed to addressing root causes of hunger and poverty. As well, the president invited me to join her in attending a three-day training on Liberation Economics led by United for a Fair Economy, a non-profit based in Boston. As these examples suggest, FBST sought influences that went beyond business models and corporate-anti-hunger alliances, as described by Fisher (2017), to include social movement organizations, innovative food banks and anti-hunger and anti-poverty organizations.

With these new models, our inquiry began to focus more about clients' life experiences and their experiences at food pantries. In Summer 2015, FBST sponsored a series of focus groups with food access program users from each county in their service area. They asked participants how they feed their families; the biggest problems they face; main reasons why they and their communities don't have enough food; and the advice they would give to policy makers to change policy (FBST Community Focus Group Facilitator's Guide 2015). In eight focus groups, they engaged nearly 80 participants "in a discussion about their lives and the challenges they face in getting by on limited or low incomes" (FBST 2015). Key staff began to notice and discuss the contradiction that, as one explained, "We had never systematically listened to our clients. That's ridiculous!"

By 2016, our inquiry team and a core FBST team was listening more to clients and drawing attention to the contradiction between providing food and the indignity of pantry experiences. FBST created a new position and hired an Advocacy and Education manager, who brought experience in Action Research and international development and began to focus on leadership development and organizational change. She worked to develop what became known as the Speaker's Bureau. Pantry users and participants in food access programs were selected by application for a nine-week leadership program, including "training in advocacy, food bank operations, public speaking, conflict resolution, and general education about the root causes of hunger" (Speakers' Bureau 2016, p. 3). At a graduation ceremony in June, seven participants shared their stories "about their journey from pantry patron to food justice advocate" (p. 3).

The Speaker's Bureau became a learning experience as much for FBST staff as for the participants. As a result, the core staff began to build more significant relationships with people with lived experience in poverty. Activities increasingly focused on cultivating leadership and on changing narratives about people experiencing hunger and poverty. A lively Facebook group became a medium through which Speaker's Bureau graduates communicated regularly despite rural isolation.

At the same time, FBST guided college students in conducting participant observation at pantries and interviews with pantry volunteers and coordinators, to learn about attitudes and behaviors toward food pantry users. After student research showed that attitudes ranged from contempt to compassion, our group also began to inquire about the educational process required to change pantry volunteers' attitudes and behaviors.

Phase 3

In the most recent phase, our learning and research continued as the advocacy program staff, college students, and SB participants studied political-economic trends in New York State, interviewed food pantry volunteers and conducted participant observation at food pantries. We deepened relationships with a national network of poor people's organizations that offers ethical principles, sociological analysis and reference points for our project. Already in 2016, FBST had signed onto the national Poor People's Campaign, referencing the campaign initiated in 1967 by Rev. Dr. Martin Luther King, Jr. just prior to his assassination (Poor People's Campaign 2016). The Campaign calls for organizations to build a "fusion politics" based in "nonviolent moral direct action" and lifting up the voices of those most directly impacted by interlocking oppressions (Poor People's Campaign 2018).

Building on this learning, the organization has incorporated lessons from our research into its strategic plan. We identified the need to address pantry volunteers' commitment to a charity model and internalized disempowerment among participants in food access programs.

By this phase, FBST had become acknowledged nationally among the Closing the Hunger Gap network (CTHG). The president explained, "CTHG said, if we need to solve hunger, we need to look at root causes." With this network, FBST has created opportunities to use its emerging justice platform to shift discourse nationally.

In 2017, the Speaker's Bureau completed its second year. FBST invited and paid for SB participants to travel to Albany for a legislative lobby day, and to the Anti-Hunger Policy Conference in Washington DC. At the Policy Conference, the graduates received a standing ovation for their presentation (Rogers 2017). In October 2017, the Poor People's Campaign chose Binghamton, NY, as the location for their mass meeting, acknowledging FBST's work with "people with lived experience in poverty." Twelve Speaker's Bureau graduates and three FBST staff led a workshop called "Pass the Mic: Storytelling and Leadership of the Poor" for the Faith for a Fair New York conference, organized by the Labor-Religion Coalition as part of building the Poor People's Campaign in New York State.

Agency leaders also encouraged both "generations" of SB graduates to seek opportunities and interact with each other. In these interactions, they noticed examples of solidarity and mutual support, as well as conflict, jealousy, competition, and even frustrations with the limits of the program. While at first concerned about the reputation of the program, agency leaders realized that these conflicts were an expression of the distresses the women experienced as a result of their economic precariousness.

As a result of the presentations at the Anti-Hunger Policy Conference, FBST has begun to receive inquiries from Food Banks across the country about how to begin Speaker's Bureau programs and how learn from FBST's efforts (Quackenbush, personal communication, May 10, 2017). In addition, in 2017, FBST was recognized by Feeding America as the Food Bank of the Year (Food Bank of the Southern Tier 2017), which may also strengthen its platform.

Discussion and analysis of changes in organizational practices from 2008 to 2018

To interpret the above findings, I analyzed written reflections by core team members and detailed notes from a core team meeting that focused on analysis. Core team members sent written reflections via email in April 2017 and in June 2017 and are used by permission. To analyze these, I applied the following categories, adapted from Fisher (2017): (1) changes in problem definition, whom the organization serves and outcomes measured, (2) increased funding for advocacy, and (3) impacts of its advocacy and education on stakeholders. I then shared this analysis with the core team and incorporated their feedback during 2018. While analysis in conventional research is typically written by in objective, reporting language, this section incorporates quotes from the core team so as to acknowledge the collaborative character of analysis in an AR process.

Defining the problem and its solutions

The core team discussed the changes in the organization's definition of hunger over the last 16 years. The organization shifted from being "inward-focused, focused on program growth and development, fundraising, policies and procedures, becoming a "best in class" food bank, etc." to being "more outward-facing, i.e. focused on the problem of hunger and food insecurity, how it manifests in our community and how we can leverage our resources most effectively to address it (email communication, April 21, 2017). The shift from inward to outward orientation reflects a shift from addressing hunger as a technical problem of food distribution to one that requires social change.

Core team members described a change in the organization's role and its definition of its stakeholders. In the past, its role was to be in service of the pantries, food programs, and other agencies that distributed food.

> "In the past, our clients were our member agencies and we thought about how to provide them with better services. Today, our clients are those in need of food and we're asking ourselves different questions" (email communication, April 21, 2017).

> "It used to be that the emergency food system was organized around the needs of volunteers. What we've been doing is attempting to organize it around the needs of the people we serve" (Meeting, June 15 2018).

A pivotal change was when staff and leaders began to focus on individuals who need food. A key ingredient in this change was that these leaders committed to treating people experiencing poverty as people with dignity, knowledge from experience, and rights.

Today, the organization measures its social impact instead of exclusively focusing on pounds of food distributed. While the FBST still tracks pounds of food distributed, number of volunteers, requests for food, and dollars raised, they have added new outcomes such as "debunking stereotypes about hunger and food insecurity" and "engaging stakeholders' heads, hands and hearts," which are admittedly "not as easy to measure and track."

Core team members sometimes express criticism of other service providers. They describe false explanations of hunger and poverty held by volunteers or service providers, such as "if only people still knew how to cook," or poor people "should just get a job." They express criticism of FBST's prior approach to service.

> "The food bank used to be like, 'Here are the scraps from our broken food system,' and they'd justify themselves by saying, 'if you're hungry, you'll eat anything." (Meeting, June 15, 2018).

They also recognize obstacles for other food banks, acknowledging why food banks stick with the "old model" of focusing on efficient food distribution and the needs of volunteers rather than the need for food.

> "The old model can be very self-serving to food banks. Food banks are well-resourced and well-respected... Lots of food banks don't want to make waves because it's going to impact their funding" (Meeting, June 15, 2018).

An important aspect of problem definition for the core team surrounds their beliefs and values about hunger, charity, and justice. In July 2018, the core team discussed a title for this paper. One draft title included the phrase "from charity to justice." In discussion, we realized this language was the title of the Closing the Hunger Gap conference which staff had attended. But for some, leaving charity behind entirely seemed unrealistic. Instead, a team member stated, "we like to use the two feet model from Catholic Social Teaching which says that both charity and justice are necessary for change." The two feet model critiques efforts that rely only on charity, but acknowledges that both addressing systemic, root causes of problems and short-term emergency assistance are necessary (United States Conference of Bishops 2018). Referring to Catholic Social Teachings offers legitimacy for the food bank, as the organization is an agency of Catholic Charities.

Members of the core team identify with and are proud of the recent changes and see themselves as leaders responsible to inform and motivate others. "Now we know we have to get everybody to the table." They acknowledge that feeding the hungry can be a way to pull people into deeper conversations. "How do we meet them where they're at, gently bring them along the path to recognize that distributing food is not the only way?" "If we get youth involved, we'd be building young minds… they could grow up to be anti-hunger advocates" (Meeting, June 15, 2018).

While the core team notices differences across the food bank's departments in terms of definitions of the problem and its solutions, there are indications of a cultural shift, of both discourse and practices (Dodd and Nelson 2018). "We're also seeing changes within the agency, finally seeping into other departments… We're spreading ideas, learning and deepening the work within the organization" (Meeting, June 15, 2018). When hunger education began in 2009, the president justified youth hunger education to the board of directors primarily because parents donated when their kids learned about hunger. But now the board understands the importance of advocacy for its own sake. A staff member explains, her co-workers do not bring together women living in poverty (the SB participants) just in order to get funding, although some funders do appreciate the value of the SB initiative. Instead, they "live the mission… Everything we do has the client at the center; this has permeated our culture" (personal communication, June 2018).

Conclusions

Reflections about collaborative inquiry and AR

In evaluation, there are strengths and limitations of the collaborative inquiry with stakeholders from a regional food bank. In listing benefits of the experience, core team members appreciated the opportunity to document, notice, reflect on the organization's changes (Meeting, July 2018). Both core team participants and SB members appreciated the involvement of people with lived experience in poverty along with professionals with formal organizational power. SB participants noticed that their knowledge was valued and could contribute meaningfully to improving the work of the food bank. In reflections on their involvement in AR research with sociology students, SB members appreciated that their experiences were valued; they were glad to practice research skills and appreciated contributing to knowledge creation. Individual participants appreciated the opportunity to contribute to FBST's practices.

Limitations of this project include both the depth and breadth of participation. The core team, including me, originally assumed, based on our knowledge of Action Research, that participants should be involved in all phases of research. However, this turned out to be ambitious and not always best for participants. As Cornwall (2008) acknowledges, no matter how participatory they seem, participants sometimes regard spaces that they are invited to as means to gain benefits or improve their own access to services, which is very understandable. SB members described contributing to changes in the Food Bank as simultaneously exciting, inspiring, empowering and exhausting, due to the fatigue of survival, parenting, and navigating social services. It is critical for service-providers who plan participatory spaces to be aware of these limitations. Rather than assume that greater participation would be better for all members of the team, Cornwall's (2008) concept of optimal participation was useful. As our collaborative inquiry continues, it will be important to assess participation according to this more nuanced measurement instead of assuming a linear progression in participation.

Conducting collaborative inquiry as a community-campus partnership also created space for sharing knowledge and reflecting among professionals and between service providers and those who receive services. Although relationships of trust and collaboration were built, there should be no pretense that this process leveled social hierarchies. Discomfort among service providers with social power differences evidences their willingness to question power relationships and unequal material conditions but does not overcome these inequalities. Some community center food banks models and social movement organizations might model social relations of dignity as they strive to address structural inequalities.

Reflections on organizational change

Program participants, leadership and staff have contributed to change the organization, from vision and strategic plan to implementation. The FBST's aspiration to interact with a new epistemic community is evidenced by the leaders and staff whose networks extend beyond the dominant anti-hunger networks, and by participation of SB participants in these networks. There is evidence of openness to new concepts and paradigms and to ongoing self-critique. Leaders, staff and program participants have sought opportunities to learn from food banks as well as social movement organizations and networks around the country and take risks to implement new practices and repertoires. The organization has added a substantial set of new activities, including listening to participants, educational workshops, leadership development training, and collaborative inquiry.

As we have documented, ideas and practices, including the concept of poverty as structural and the commitment of listening to those who need food, have spread beyond the Advocacy and Education department. Across its departments, the organization has built a new commitment to listen to people living in poverty who need food. Resources dedicated to advocacy and education are substantial and challenge staff to develop new skills. As the organization envisions its role as leading other organizations in the region in the effort to end hunger and poverty, it is developing new repertoires of action. There are also indications of a new organizational compass which goes beyond the corporate model, with values of dignity, respect and mutual care articulated in its strategic plans and enacted in SB meetings.

Although the program is new, and staff acknowledge the need for further evaluation, participants offer preliminary evidence that the Speaker's Bureau has had positive impacts on their self-esteem. Participant have challenged internalized shame, changed their understandings of the causes of poverty, and gained motivation and opportunities for leadership. Future efforts should aim to mitigate participation fatigue and develop organizational practices that support genuine involvement and opportunities. As our collaborative inquiry continues, subsequent steps will be to document impacts of the SB on staff, board members, donors and others.

From organizational change and community development perspectives, these changes may be cautiously interpreted as a step beyond a charity model. From a social movement perspective, for FBST to be a leader in a regional movement to end hunger, as its latest strategic plan suggests, will require its impact to exceed its own organizational boundaries. Further implementation will be necessary to educate the public about the roots of hunger in structural poverty, to provide leadership among anti-hunger organizations, and to build support in new communities in the region. As dispossession continues in New York's southern tier, FBST may do well to connect with social movement organizations in the state and with national networks, such as by expanding involvement with the Poor People's Campaign and the broader movement to end poverty, which recognize that structural problems such as poverty are unassailable if they are addressed piecemeal.

Despite FBST's achievements, deep concerns regarding the emergency food system remain. The organization should engage in ongoing reflection including studying critiques of the emergency food regime, including Poppendieck (1999), Fisher (2017), and Winne (2008). Organizational self-study could include critical questions such as these: How can the organization get out of the cycle of growth and contribute to significantly reducing food distribution, or "shortening the line"? How can it overcome the influence of its capital campaign and its building structuring its activities? How can it avoid tendencies of corporate culture that prevent risk-taking and innovation? How can the organization redefine its corporate partnerships so that corporate priorities do not limit its approach to ending poverty and hunger? Can FBST continue to shift resources toward broad-based efforts to end poverty?

A fundamental question is of the relationship between charity and justice. In emphasizing the Catholic Social Teachings, FBST acknowledges that coordinating the "two feet" of charity and justice requires discussion about morality, values and politics. It requires challenging power dynamics both within and beyond the organization. Expanding their epistemic communities may offer additional references and innovative models, such as the Black Panther Party's free breakfast distribution, survival projects within the movement to end poverty (Baptist and Rehmann 2011), and the human right to food (Chilton and Rose 2009).

Implications

This project offers empirical evidence that food regimes in general and the emergency food regime in particular, are indeed dynamic constellations. In focusing on food banks, we acknowledge constraints, including their entrenchment in relationships with corporate funding and their tendency to provide a moral safety valve, but hold out the possibility that they might provide leadership and resources that can go beyond charity. As Winne (2008) acknowledges, leaders of food banks are particularly well positioned to promote "a vital public discourse around hunger, food insecurity and poverty" (p. 76).

Increasing evidence suggests (Dodd and Nelson 2018; Galinson 2018) that it may be possible for food banks to shift their discourse and practices. In this case, a collaborative inquiry, guided by principles of AR has involved food bank leaders and staff, people living in poverty and college students, in asking and answering questions about how a food bank can move beyond charity. In this process, FBST demonstrates characteristics of a learning organization (McNiff and Whitehead 2000). As our collaboration continues, we join critics of charity food banks in acknowledging the impact food banks may have if they put their authority, lived experience and networks to the task of building a movement to end hunger and poverty.

Acknowledgements I gratefully acknowledge the other members of the core team for their commitment and feedback throughout the research process. I am grateful for the energetic involvement of participants in FBST's Speaker's Bureau, students from my Inquiry and Action class, and graduate assistant Rachelle Sartori, who contributed to the development of portions of this research. I also thank the core team, Tim Shenk, Diane Swords and Pamela Sertzen for feedback on drafts. Portions of this research were funded by the Ithaca College Center for Faculty Excellence and the Ithaca College Center for Civic Engagement.

References

Atkins, P., P. Blumenthal, A. Edisis, A. Friedhoff, L. Curran, L. Lowry, L. Travis, T. St. Clair, H. Wial, and H. Wolman. 2011. *Responding to manufacturing job loss: What can economic development policy do?*. Washington, DC: The Brookings Institution.

Attygale, L. 2017. The context experts. *Tamarack Institute*. https://www.aohc.org/sites/default/files/documents/Tamarack_The%20Context%20Experts.pdf. Accessed 20 June 2018.

Baptist, W., and J. Rehmann. 2011. *Pedagogy of the poor: Building the movement to end poverty*. New York, NY: Teachers College Press.

Bluestone, B., and B. Harrison. 1982. *The deindustrialization of America: Plant closings, community abandonment, and the dismantling of basic industry*. New York, NY: Basic Books.

Chilton, M., and D. Rose. 2009. A rights-based approach to food insecurity in the United States. *American Journal of Public Health* 99 (7): 1203–1211.

Coleman-Jensen, A., M.P. Rabbitt, C.A. Gregory, and A. Singh. 2017. Household food security in the United States in 2016, ERR-237, U.S. Department of Agriculture, Economic Research Service. https://www.ers.usda.gov/publications/pub-details/?pubid=84972. Accessed 5 Nov 2017.

Collins, S.B. 2005. An understanding of poverty from those who are poor. *Action Research* 3 (1): 9–31. https://doi.org/10.1177/1476750305047983.

Cornwall, A. 2008. Unpacking 'participation': Models, meanings and practices. *Community Development Journal* 43 (3): 269–283.

DeVault, M.L. 1990. Talking and listening from women's standpoint: Feminist strategies for interviewing and analysis. *Social Problems* 37 (1): 96–116.

DiNapoli, T.P. 2010. The changing manufacturing sector in upstate New York: Opportunities for growth. https://www.osc.state.ny.us/localgov/pubs/research/manufacturingreport.pdf. Accessed 28 April 2017.

DiNapoli, T.P. 2013. Employment trends in New York State. https://www.osc.state.ny.us/reports/economic/employment_trends_nys_2013.pdf. Accessed 20 July 2018.

DiNapoli, T.P. 2015. New York State Employment Trends. https://www.osc.state.ny.us/reports/economic/employment_trends_nys_2015.pdf. Accessed 20 July 2018.

Dodd, W., and E. Nelson. 2018. Shifting discourse and practice on food banks: Insights from a community–university partnership. *VOLUNTAS: International Journal of Voluntary and Nonprofit Organizations* 1: 1–13. https://doi.org/10.1007/s11266-018-0012-0.

Esterberg, K.G. 2002. Narrative analysis. Ch. 9 in *Qualitative methods in social research*. Boston, MA: McGraw Hill, 181–197.

Feeding America. 2016. Annual report. http://www.feedingamerica.org/about-us/about-feeding-america/annual-report/2016-feeding-america-annual-report.pdf. Accessed 26 April 2017.

Feeding America. 2017. Our leadership. Feeding America. http://www.feedingamerica.org/about-us/leadership/. Accessed 26 April 2017.

Feeding America. 2008. Annual report. Feeding America. http://www.feedingamerica.org/about-us/about-feeding-america/annual-report/2008-feeding-america-annual-report.pdf. Accessed 26 April 2017.

Fisher, A. 2017. *Big hunger: The unholy alliance between corporate America and anti-hunger groups*. Cambridge, MA: MIT Press.

Fitchen, J.M. 1981. *Poverty in rural America: A case study*. Boulder, CO: Westview Press.

Fitchen, J.M. 1991. *Endangered spaces, enduring places: Change, identity, and survival in rural America*. Boulder, CO: Westview Press Inc.

Food Bank of the Southern Tier. 2015. *Community Focus Groups Report*. Unpublished document.

Food Bank of the Southern Tier. 2017. Food Bank of the Southern Tier named Food Bank of the Year. Press Release. https://www.foodbankst.org/usr/Food_Bank_of_the_Year/PR_Food_Bank_of_the_Year.pdf. Accessed 18 Nov 2017.

Friedmann, H. 1993. The political economy of food: A global crisis. *New Left Review* I (197): 29–57.

Friedmann, H. 2009. Discussion: Moving food regimes forward: Reflections on symposium essays. *Agriculture and Human Values* 26: 335–344.

Friedmann, H., and P. McMichael. 1989. Agriculture and the state system: The rise and decline of national agricultures, 1870 to the Present. *Sociologia Ruralis* 29 (2): 93–117.

Galinson, S.A. 2018. *From frozen turkeys to legislative wins: How food banks put advocacy on the menu* (Order No. 10810723). Available from ProQuest Dissertations & Theses Global. (2038490562). http://ezproxy.ithaca.edu:2048/login?url=https://search-proquest-com.ezproxy.ithaca.edu/docview/2038490562?accountid=11644. Accessed 20 July 2018.

Ganz, M. 2011. Public Narrative, Collective Action, and Power. In *Accountability through public opinion: From inertia to public action*, eds. Sina Odugbemi and Taeku Lee: 273–289. Washington D.C: The World Bank.

Goldsmith, W., and E. Blakely. 2010. *Separate societies: Poverty and inequality in US cities*. Philadelphia, PA: Temple University Press.

Gramsci, A. 1971. *Selections from the Prison Notebooks of Antonio Gramsci. Edited and Translated by Quintin Hoare and Geoffrey Nowell Smith*. London: International Publishers.

Greenwood, D.J., and M. Levin. 2007. *Introduction to action research: Social research for social change*. Thousand Oaks, CA: Sage Publications.

Gundersen, C., A. Dewey, A. Crumbaugh, M. Kato, and E. Engelhard. 2016. *Map the meal gap 2016: Food insecurity and child food insecurity estimates at the county level*. Feeding America.

Haas, P.M. 1992. Introduction: Epistemic communities and international policy coordination. *International Organization* 46 (1): 1–35.

Harvey, D. 2003. *The new imperialism*. New York, NY: Oxford University Press.

Kania, J., and M. Kramer. 2011. Collective impact. *Stanford Social Innovation Review*. 36-41. https://ssir.org/articles/entry/collective_impact Accessed 20 July 2018.

Kelly, J.G., L.O. Mock, and D.S. Tandon. 2001. Collaborative inquiry with African-American community leaders: Comments on a participatory action research process. In *Handbook of action research*, ed. P. Reason, and H. Bradbury, 348–355. London: Sage Publications.

Knowles, M., J. Rabinowich, T. Gaines-Turner, and M. Chilton. 2015. Witnesses to hunger: Methods for photovoice and participatory action research in public health. *Human Organization* 74 (3): 255–265.

Marx, K. [1867] 1990. *Capital, Volume I*. London: Penguin Books Limited.

McIntyre, L., D. Tougas, K. Rondeau, and C.L. Mah. 2016. 'In'-sights about food banks from a critical interpretive synthesis of the academic literature. *Agriculture and Human Values* 33 (4): 843–859.

McMichael, P. 2005. Global development and the corporate food regime. In *New directions in the sociology of global development*, ed. F.H. Buttel and P. McMichael, 265–299. Bingley: Emerald Group Publishing Limited.

McMichael, P. 2009. A food regime analysis of the 'world food crisis'. *Agriculture and Human Values* 26 (4): 281.

McNiff, J., and J. Whitehead. 2000. *Action research in organisations*. London: Routledge.

Mishel, L. 2012. Why aren't workers benefiting from labour productivity growth in the United States? *International Productivity Monitor* 23: 31–43.

Park, P. 1993. What is participatory research? A theoretical and methodological perspective. In *Voices of change; participatory research in the United States and Canada*, eds. P. Park, M. Brydon-Miller, B. Hall, and T. Jackson, 1–19. Toronto, ON: OISE Press.

Poppendieck, J. 1999. *Sweet charity? Emergency food and the end of entitlement*. New York, NY: Penguin.

Poppendieck, J. 2014. Food Assistance, Hunger and the End of Welfare in the USA. In *First world hunger revisited: Food charity or the right to food*, 2nd ed, ed. G. Riches and T. Silvasti. Basingstoke: Palgrave Macmillan.

Poor People's Campaign. 2016. Who's endorsing the Poor People's Campaign. https://poorpeoplescampaign.org. Accessed 20 April 2017.

Reinsborough, P., and D. Canning. 2010. *Re: Imagining change: How to use story-based strategy to win campaigns, build movements, and change the world*. Oakland, CA: PM Press.

Roadside Theater. 1999. *Story circle methodology*. https://roadside.org/sites/default/files/storycircle.description.web_.2012.pdf. Accessed 30 July 2018.

Robinson, W.I. 2008. *Latin America and global capitalism: A critical globalization perspective*. Baltimore, MD: The Johns Hopkins University Press.

Rogers, C. 2017. Food Bank goes to D.C. *The Advocate*. (A publication of the Food Bank of the Southern Tier.) March 2017. https://us14.campaign-archive.com/?u=d08d63f3f07d1c0240225dfe3&id=6bc6ea89cf&e=2ab1304af8#dc. Accessed 18 Nov 2017.

Scott, R. 2015. Manufacturing job loss: Trade not productivity is the culprit. *Economic Policy Institute Issue Brief #401*, https://www.epi.org/publication/manufacturing-job-loss-trade-not-productivity-is-the-culprit/. Accessed 29 July 2018.

Shaefer, H.L. and K. Edin. 2012. Extreme poverty in the United States 1996 to 2011. *National Poverty Center Policy Brief #28*. http://www.npc.umich.edu/publications/policy_briefs/brief28/policybrief28.pdf Accessed 18 Nov 2017.

Sommeiller, E., Price, M., and E. Wazeter. 2016. Income inequality in the U.S. by state, metropolitan area, and county. *Economic Policy Institute*, June 16 2016. http://www.epi.org/publication/income-inequality-in-the-us/. Accessed 16 April 2017.

"Speaker's Bureau: From pantry patrons to hunger advocates" 2016. *The Harvester: A Publication of the Food Bank of the Southern Tier.* (Fall 2016). Retrieved from https://www.foodbankst.org/usr/Harvester%20Newsletters/September_2016_Harvester_final.pdf. Accessed 18 April 2017.

St. Clair, C., and M.A. Dunlea. 2012. Hungry in New York: A 2012 study of emergency food programs in New York State. *Hunger Action Network of New York State.* http://www.hungeractionnys.org/wp/wp-content/uploads/2012/12/EFP-report-Final.pdf. Accessed 15 April 2017.

Tilly, C. 2002. *Stories, identities, and political change*. Lanham, MD: Rowman & Littlefield.

U.S. Bureau of Economic Analysis. 2017. Total gross domestic product for New York [NYNGSP]. FRED, Federal Reserve Bank of St. Louis. https://fred.stlouisfed.org/series/NYNGSP. Accessed 24 Aug 2017.

U.S. Census Bureau. 2017. How the Census Bureau measures poverty. https://www.census.gov/library/visualizations/2014/demo/poverty_measure-how.html. Accessed 15 June 2018.

United States Department of Agriculture Economic Research Service. 2013. Measurement. *Food Security in the United States.* https://www.ers.usda.gov/topics/food-nutrition-assistance/food-security-in-the-us/measurement.aspx. Accessed 2 Oct 2017.

United Way. 2016. ALICE: Asset Limited Income Constrained Employed New York study of financial hardship. https://www.unitedwayalice.org/new-york. Accessed 29 July 2018.

Valera, P., J. Gallin, D. Schuk, and N. Davis. 2009. Trying to eat healthy: A photovoice study about women's access to healthy food in New York City. *Affilia* 24 (3): 300–314.

Vitiello, D., J.A. Grisso, K.L., Whiteside, et al. 2015. From commodity surplus to food justice: Food banks and local agriculture in the United States. *Agriculture and Human Values* 32: 419.

Wang, C.C. 1999. Photovoice: A participatory action research strategy applied to women's health. *Journal of Women's Health* 8 (2): 185–192.

Winne, M. 2008. *Closing the food gap: Resetting the table in the land of plenty*. Boston, MA: Beacon Press.

Zucchino, D. 1997. *Myth of the welfare queen: A Pulitzer Prize-winning journalist's portrait of women on the line*. New York, NY: Touchstone.

Publisher's Note Springer Nature remains neutral with regard to jurisdictional claims in published maps and institutional affiliations.

Alicia Swords is associate professor of sociology at Ithaca College. Current research addresses community-based learning and social movements to end hunger and poverty.

Agriculture and Human Values (2019) 36:867–878
https://doi.org/10.1007/s10460-019-09947-w

SYMPOSIUM/SPECIAL ISSUE

Gardens and Green Spaces: placemaking and Black entrepreneurialism in Cleveland, Ohio

Justine Lindemann[1]

Accepted: 11 May 2019 / Published online: 20 May 2019

Abstract
This paper presents a case study of Gardens and Green Spaces (GGS), a resident-driven, grant-funded project in Cleveland, Ohio working toward community change. Through both placemaking and entrepreneurial strategies, the main grant objectives are to effect change at the intersection of food (and agriculture), arts, and culture in Kinsman, a 96% Black Neighborhood on Cleveland's east side. While community development (CD) projects are often designed by outside 'experts' who inform the scope and focus of grant-funded projects, this project is rooted in the hypothesis that a resident-driven approach to CD will lead to increased community buy-in and participation, resulting in more lasting and substantive community change. GGS works across sectors, integrating arts, culture, and food to promote placemaking and community-based entrepreneurial engagement as a path towards greater health, equity, and economic resilience. This paper argues that community-based and resident-driven development—although not without its own challenges—can result in more holistic community transformation across sectors, with the potential for greater resident participation, sustainability, and equity. The case study presented in this paper, including in-depth interviews and neighborhood surveys, is an examination of the pilot phase of GGS, and argues that both placemaking and entrepreneurialism represent a negotiation between market driven community development and a solely philanthropic model. It provides insight into more equitable and sustainable change that has the potential to shift the traditional paradigm of expert driven, or "outside-in" community development.

Keywords Community development · Cleveland · Urban food production · Creative placemaking · Urban agriculture · Urban planning · Land · Food · Arts · Culture · Philanthropy

Abbreviations
CD Community development
GGS Gardens and Green Spaces

Introduction

This paper presents two of the strategies—placemaking and entrepreneurialism—of the Gardens and Green Spaces (GGS) project in the Garden Valley neighborhood in Cleveland, that strives for community change at the intersection of arts, culture, food, and agriculture. This project holds as an underlying premise that transformations in an inequitable urban food system always occur within a larger dialectic of urban metabolism (Heynen et al. 2006; Smith 2008). According to this perspective (Heynen et al. 2006), urban metabolism or transformation occurs as a result of both human labor and the work of non-human natures (including the built environment), which is an important recognition of the complexity of urban agriculture, community food systems, and community development more broadly, and supports the contention that equitable food system transformation must extend beyond interventions solely around growing food.

Erika Allen, co-founder of the Urban Growers Collective in Chicago writes about a community food system that "deals with everything; all the components that are needed to establish, maintain, and perpetually sustain a civilization" (Allen 2010). This concept of "everything" includes the struggle for sovereignty around community land and water rights, but also extends as far as community participation in conversations and planning around transportation, energy, housing, economic development and beyond. Questions of labor, gender, race, health care, age, education, and social services (among others) exist alongside land rights,

✉ Justine Lindemann
jfl93@cornell.edu

[1] Development Sociology, Cornell University, Warren Hall, Ithaca, NY 14853, USA

ecological sustainability, food access, nutrition, and sovereignty as important components of a community development intervention whose ultimate goal is to effectively and sustainably reduce hunger. Through this lens, urban food production is one piece—albeit an important one—in producing more equitable community infrastructures, healing historical traumas from economic, political, and spatial marginalization, and in establishing community resilience. While urban agriculture has received significant attention over the past several years as a strategy for economic resilience and access to increased nutrition (Grewal and Grewal 2012; Kaufman and Bailkey 2000; Sommers and Smit 1994), especially in low-income neighborhoods and neighborhoods of color (Tornaghi 2017), it cannot be the only point of intervention in sustainable and equitable approaches to community development. Sustainable and equitable community change might start with food as a powerful entry point, but must strive to encompass most (if not all) aspects of the lives of community members, engaging them not as recipients of a nutritional or health intervention, but rather as multifaceted and complex human beings.

Throughout interviews and conversations with Black urban growers in Cleveland, the importance of a holistic approach to community development was repeatedly articulated. Epistemically, participants understand food and growing as a part of the larger dialectic of transformation (Heynen et al. 2006), with important, mutually constitutive foci including other facets of their lives: education, transportation, art, literature, poetry, exercise, safety and violence reduction, and spirituality or religion. The Garden Valley neighborhood faces challenges across many different aspects of life, and as one research participant explained, community development and change must address all of these if it is to catalyze sustainable improvements in any area.

The theory of change that emerged from GGS emphasized creative placemaking and entrepreneurialism as catalytic to community transformation. The project planning team understands placemaking as a celebration and expression of culture unique to a community that strengthens, empowers, and 'activates' spaces within that community. Placemaking, as an approach to community change, relies upon already-existing assets, values, and knowledge to foster health, safety, prosperity, and well-being (Montgomery 2016) that might be overlooked through a traditional economic or market-based lens. Following Reese (2018), the project's attention to the geographies of Black food and land through a lens of what is, rather than what isn't, allows for "(b)lack ways of being, knowing, and doing" to become the drivers behind community change (p. 408). While placemaking is not an entirely new approach to achieving community change, community development projects rarely place the emphasis on holistic community transformation. Placemaking is often either at the periphery of CD practices in Cleveland, or implemented in such a way that it does not meaningfully engage with community knowledge about problems impacting (or potential solutions within) the community, or effectively acknowledge the particular histories of place.

As part of a strategy for community change, entrepreneurship can easily be critiqued as reinforcing neoliberal strategies of governance, especially those condoning the retreat of social services and state investment in particular neighborhoods (Derickson 2014; Brenner and Theodore 2002). Community residents consistently spoke about entrepreneurialism not as an individualist approach to economic gain or a replacement for state investment, however, but as economic engagement for community benefit. In other words, entrepreneurial engagement in Black communities has the potential to be an important complement to strategies of placemaking. An intentionally community-based approach to entrepreneurialism can be seen as a powerful tool for equity and sustainability in community development, especially in disinvested communities of color where traditional economic interactions have historically been oppressive and exploitative.

Literature review: community development, placemaking, and entrepreneurialism

Engaging with poor and/or historically marginalized people as full, complex, and multi-faceted human beings with needs, interests, or struggles beyond food, jobs, or shelter—that is, engaging holistically—is often not possible in community development grant projects with singular goals, such as combatting obesity or improving literacy. The ideal of holistic engagement rarely manifests in the methods deployed to address challenges facing underserved neighborhoods (see, for example, Cavanaugh et al. 2014; Cummins et al. 2014). Community development, social welfare, and government assistance programs often approach poor communities of color from a 'needs only' perspective, with a focus on survival necessities (O'Brien et al. 2004; Shannon 2014). To wit, many so-called diet-related health problems are discursively delimited to impoverished neighborhoods, especially neighborhoods of color (Shannon 2014), with an emphasis on the ways in which neighborhood environmental factors may contribute to negative social, economic, and health outcomes (Stokols 1995). These neighborhoods and communities become geographically bounded areas for intervention around focal points including food, housing, healthcare, or employment (Morales 2009). This single-issue model of intervention often uses market-oriented rather than community-based tools, disregarding cultural aspects of a neighborhood and the holistic or cross-sectional nature of community needs. The arts—and culture or creative expression more broadly—are implicitly understood to be secondarily important or even excessive and unnecessary, despite evidence that participation in the arts leads to more engaged residents and more resilient communities (Kay 2000; Purcell 2009).

Food is an excellent entry point—a universal human need—to much more complex facets of the human experience, including creativity, spirituality, and community empowerment. Growing food allows humans to articulate together with non-human nature in a way that can produce more equitable, healthy, sustainable, and just urban natures (Heynen et al. 2012). Food production is also an effective introduction to histories and geographies that have contributed to particular urban environments, and has underserved marginalized populations, not allowing them a voice in political, economic, social, and spatial processes (Heynen et al. 2006.)

Urban agriculture produced through Black labor, as an approach to a more equitable and resilient community food system (Porter 2018), challenges dominant conceptions about what 'Black' space looks like (Ekers and Loftus 2012; Harvey 2003), while also contesting power dynamics around uneven development and land use practices. Black Americans have suffered disproportionate land loss (Mitchell 2005; Pennick 1990), theft, and the fracturing of communities through policies of redevelopment and urban renewal (Michney 2011; Pritchett 2003). Within the urban context, food production often exists as a contingent form of land use, often emerging as a temporary solution to economic downturn or increases in demand for inexpensive food (Saldivar-Tanaka and Krasny 2004). According to city planners, it does not always represent the "highest and best use" of land (Lawson 2004; Németh and Langhorst 2014), and while long-term urban agricultural projects tend to do best in neighborhoods with little to no development pressure, investment in these neighborhoods is likely to be very low. What are often assumed to be either 'natural' or unintentional patterns in urban development across space and time emerge as the products of powerful socio-spatial forces; growers—and, in the context of GGS, artists and culture-bearers—labor to transform and (re)produce the spaces around them, and in so doing, shine a light on these historical–geographical inequities (Harvey 2003; Safransky 2017; Smith 2008).

For the purposes of this article, it is important to distinguish between space and place. Within many disciplines, space is theorized as a location without social connection (Tuan 1977), while Lefebvre (1996), Swyngedouw (1996), and others theorize that space as *inherently* social, pushing back against the "blank slate" theory of urban planning, development, and change (Brenner and Theodore 2002; Lefebvre 1996; Scott 2010). Theories of placemaking define place as socially produced areas—"temporary permanences" (Harvey 1996)—built to respond to the needs of residents. For the purposes of this paper, 'place' is used to refer to locations targeted by placemaking endeavors. However, in keeping with the literature (Ekers and Loftus 2012; Heynen et al. 2006), I use space to refer more generally to localities, and contend that no space—or place—is devoid of social relations or removed from political decision-making processes. Lastly, Lefebvre's writing on "the production of space" is useful in explaining processes of community and creative placemaking. This work focuses on the inherently social nature of space, its production through both human labor and nonhuman transformation, and, importantly, the relationship between (urban) space, the state, and economic relations (Lefebvre 1991, 1996, 2009).

Placemaking and entrepreneurialism

Creative placemaking projects often encourage entrepreneurialism as a way to fill an economic vacuum in otherwise economically and spatially marginalized neighborhoods. Ideologies of entrepreneurialism and market-based growth are deeply ingrained in urban development practices, even in communities who have not been served by this model. Entrepreneurial ideologies often encourage residents to become self-sustaining and to produce their own means of subsistence. Especially in low-income neighborhoods, entrepreneurs rarely receive the tax subsidies or abatements that larger-scale businesses do in cities looking to revitalize (Wilson 2007). Smaller-scale entrepreneurs also do not benefit from economies of scale and yet, are expected to become self-sustaining within a relatively short timeframe.

Black entrepreneurship, however, was and is necessary in the face of a state and economy that have neglected, oppressed, and excluded Black communities from economic, political, and social success and power. Within the Black community, entrepreneurial endeavors are often linked to the historical struggles and traumas experienced by Black Americans. Enterprises that address the violent and oppressive history of Black Americans, while simultaneously celebrating Black culture, positions Black history and culture to heal trauma and promote community solidarity. Learning about and reframing historical traumas is a key part of many placemaking projects within Black communities (Sbicca 2012), and is essential to building stronger communities and stimulating community economies.

In marginalized communities, entrepreneurs emerged out of a need to build amenities such as medical offices, restaurants, and lodgings, contributing to a uniquely Black class consciousness amongst Black entrepreneurs. Black Wall Street and the Harlem Renaissance represent successful, albeit time-bound, examples of Black entrepreneurship that supported the creation of local economies while maintaining control over intellectual property and, at the same time, building community power. Arts and culture were central to these entrepreneurial endeavors, as was placemaking and a rootedness in place. Artists generated income and became upwardly mobile without relying on government support. These examples represent ideologies of entrepreneurship based not in individual economic achievement but rather in community empowerment, bolstering Black economies, and in resisting dependence on a racially oppressive state that, over time, has disinvested from, segregated, fractured, and disproportionately incarcerated Black communities (Brahinsky 2013; Massey and Fischer 2000). Black entrepreneurialism simultaneously reinforces the importance of market-based capitalist growth while also building alternative community economies—often communities of care (Sundin 2011)—and Black resilience in the face of continued neglect by traditional investment streams. Much like placemaking, Black entrepreneurialism in practice does not guarantee a reversal of hegemonic neoliberal tendencies across urban space nor does it promise greater equity in investment or community development. I present these arguments as a way to dismantle the idea that any hegemonic practice or ideology exists without counterhegemonic opposition.

Placemaking and philanthropic capital

Differential investment in and approaches to placemaking, philanthropy, and CD across space often correlate to race and class (Ley 2003; Montgomery 2016). Whereas downtown areas and more affluent neighborhoods see considerable increases in investment in and demands on housing, poorer neighborhoods struggle to recover not only from economic decline and the foreclosure crisis of 2007–2008, but from decades of redlining, disinvestment, and institutionalized discrimination. As a part of community and urban development strategies, creative placemaking attempts to disrupt these racialized patterns, expanding beyond interventions that target a particular problem or behavior to instead 'activate' spaces, and bolster community culture and economies in a more holistic sense. This happens by bringing together networks of people to socially appropriate space (Lefebvre 1996, 2009), changing its social and spatial dynamics, and infusing community significance to that space. Placemaking is one way to produce space that has the potential to better reflect the needs, desires, and cultural histories of community members (Bain and Landau 2017).

Philanthropic dollars allow for creative and collaborative placemaking to increase the artistic and cultural draw of a neighborhood or community, and engage community residents in the creation and enactment of their vision for space. Municipal or state investment in cultural infrastructure tends to focus on more affluent neighborhoods, and is larger in scale. Kinsman (and Garden Valley, which is part of Kinsman) have historically received little economic investment from the state, and most investment in the arts or in cultural infrastructure stems from grants and philanthropic dollars rather than local or state-level subsidies. Philanthropic support of placemaking also has the potential to contest or partially displace a neoliberalized privileging of economic growth to the detriment of social relations. This supports a more expansive and holistic paradigm of neighborhood change, and a production of space that allows for alternative visions of the urban to be enacted from a grassroots perspective (Angelo 2017; Lefebvre 1991).

While placemaking can be seen as resistance against homogenized (and thus commodifiable) urban space (Lefebvre 1991, 1996), it is far from a panacea. Placemaking is not able to fully contest a market-dominant and globalized approach to urban development (Jessop 2000; Sassen 2000; While et al. 2004). Nor does it truly shift patterns of economic investment at either the local or the global scale. Indeed, support by community foundations and other grantors have done little to reverse decades of economic and political marginalization. Further, placemaking, like entrepreneurialism, can contribute to the market fundamentalist paradigm of development, as residents and grantees risk perpetual reliance upon local market growth and the continued solicitation of grant funding as a way to both prolong investment in place and attempt to equalize uneven allocations of capital across space. Placemaking thus risks reinforcing neoliberal tenets of capitalist development at the community scale rather than an actual shift in the power dynamics of urban planning or urban imaginaries. Economic sustainability and equity must be built into the structure of any project: resident participation, plans for economic viability in the long-term, and continual evaluation on progress towards these goals.

Historical geographical background

Cleveland follows many other North American cities, with both austerity politics and investment reifying patterns of marginalization and uneven development, often ushering in or hastening the gentrification of poor neighborhoods of color as spaces of commodification, and reifying market ideologies of development across urban space (Ley 2003; Porter 1995). The Cleveland Museum of Art is located in University Circle, the wealthiest neighborhood in Cleveland; three professional sports arenas sit in a revitalized downtown business district, an area of heavy investment and affluence; rapidly gentrifying neighborhoods such as Collinwood and the Detroit Shoreway have seen significant investment in the arts and business that coincides with rising real estate and rental costs. While gentrification is not the foregone conclusion to investment in place, the literature suggests a strong correlation between investment in creative institutions, and affluence in urban space (Grodach and Loukaitou-Sideris 2007). Furthermore, development in one neighborhood can have very real impacts on neighboring communities that do not receive the same levels of investment (Lindemann 2019). Case Western Reserve University in University Circle owns dozens of properties in neighboring low-income communities of color where real estate costs are lower, in effect decreasing available housing stock for those community residents while simultaneously providing amenities for incoming wealthier occupants (Richmond 2017).

Gardens and Green Spaces takes place in the Garden Valley neighborhood on Cleveland's east side, sometimes called the Forgotten Triangle. Garden Valley is located in the larger Kinsman neighborhood of Cleveland's Ward 5. It is also just east of the Central neighborhood, where the majority of Black migrants to Cleveland settled during the Great Migration, and where the segregated Black-only public housing, Outhwaite Homes, was built in the 1930s. This community has some of the highest poverty indicators in Cleveland, including female-headed households, high rates of incarceration, unemployment (almost 30%), and low educational attainment, as well as poor health indicators. Approximately 96% of residents are African-American, and as of 2013 had a median household income of about $14,000. About 84% of the housing is rental, and 35% of Cuyahoga County's public housing (over 4000 units) is concentrated in and around the neighborhood. Garden Valley is one of the original sites of federally funded urban renewal, with several hundred apartment units built in the 1960s (Michney 2011). Designed to attract middle-income Black residents, housing projects in Garden Valley instead became an example of how racial segregation and the concentration of poverty were simultaneously reinforced through the enactment of urban renewal policies (Jenkins 2001; Michney 2011). While parts of Cleveland now boast an economic renaissance and population growth, population loss continues in Kinsman, especially in the wake of the foreclosure crisis, which catalyzed an 84 percent decrease in housing values in the Kinsman neighborhood (WRLC 2015).

Statistical indicators, while only part of this community's story, are significant to the ways in which development projects have approached this community, with each indicator representing a unique target for community development investment or inquiry. At the city scale, investment in large-scale development tends to eclipse funding for alternative approaches such as urban agriculture. Kinsman and Garden Valley have incredibly high concentrations of vacant land in Cleveland (WRLC 2015), representing an opportunity to reimagine both space in the abstract and more concrete urban and community development initiatives (Fig. 1).

Beginning in the mid-2000s a group of three childhood friends and former residents began cleaning up about two acres of vacant land, in part through extensive soil remediation. They founded the Rid-All Green Partnership Farm, one of the core partners in GGS. Rid-All (now spanning more than seven acres) has become an example of a successful and profitable urban farming venture with a focus on social justice and neighborhood engagement. While Rid-All represents hope for the positive impact of alternative land use projects, the surrounding community has not recovered from decades of misuse, neglect, and the resulting socio-spatial challenges. On the one hand, Kinsman sits at the forefront of alternative land use innovation in Cleveland, with the support of the local Community Development Corporation (CDC), the Ohio State University Extension, and zoning ordinances established through the City Planning Commission that allow food production and market stands in residentially zoned areas. Many agriculture projects in this community span several acres of land. Community gardens, an incubator farm, an aquaculture business, a 3-acre greenhouse production site, as well as other smaller projects are all situated in Kinsman. A 28-acre "Urban Agriculture Innovation Zone" is intended for the transformation of fallow land into sites of food production. Residents interested in farming have the opportunity to rent a plot on the incubator farm to hone their skills while accessing a support system of extension agents and farmers. Several community gardens on previously vacant land also serve as sites for community gatherings and horticultural education. The area's focus on local food has increased fresh food production, providing some residents with hyperlocal and healthy eating options in an area that is intensely impacted by "food apartheid" (Holt-Giménez and Wang 2012).

Notwithstanding, traditional development investment and the associated transformations do not always touch down to impact residents' quality of life. The Opportunity Corridor, a multi-million-dollar transportation project, will connect a nearby highway interchange with the affluent University Circle area (OHDOT). Funding comes from both local and state sources, and, at a community level, has contributed to some street-front improvement on Kinsman Avenue (intersecting Garden Valley). These improvements have brought a few businesses (most of which were founded by the local CDC) into the neighborhood but without significant economic opportunity. Residents do not often frequent these businesses due to barriers both perceived (a feeling of not belonging) and tangible (prohibitive cost); rather, they mostly serve commuters who work in the community development or construction sectors. The Opportunity Corridor is often described as a new iteration of urban renewal, with eminent domain claiming land for current or future development, (re)producing Kinsman as a community to pass through rather than one in which to spend time or money (see Michney 2011).

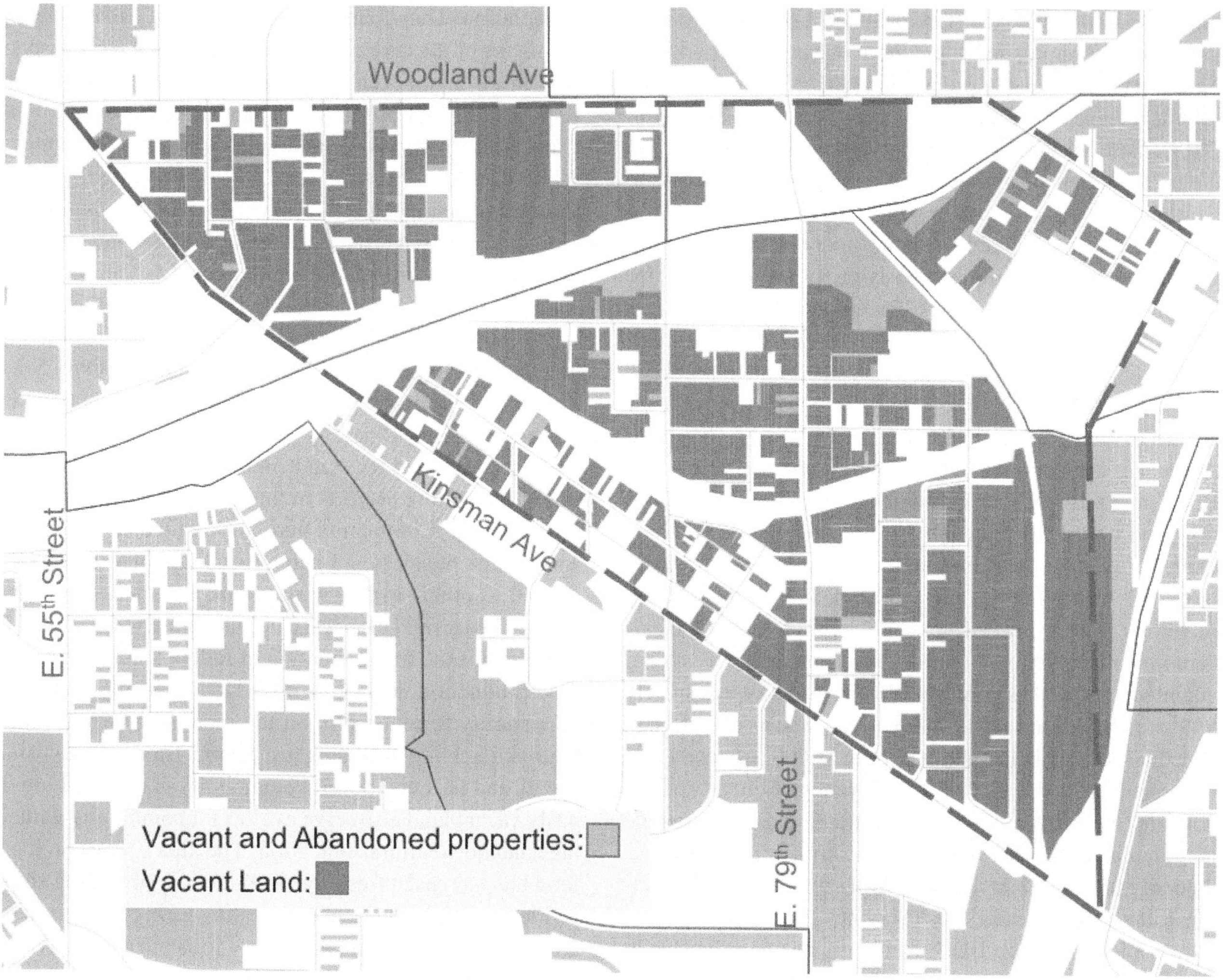

Fig. 1 Vacant properties and land in the forgotten triangle. Prepared by author using 2010 Census data and 2015 property survey data from the Western Reserve Land Conservancy, housed at NEOCANDO

The gardens and green spaces pilot program

The lived experiences of planning team members lends a familiarity with the history of the Garden Valley neighborhood and the struggles facing residents in their everyday lives. It is with this deep awareness that GGS began with food—growing, cooking, sharing, and eating food—and expanded from this platform to incorporate music, arts, and other cultural engagements within the community. Recognizing both the centrality of the many challenges facing the neighborhood and its residents, and the necessity to build community resilience (and community economies) from within, GGS was constituted as a partnership between four Black-run and owned organizations and businesses. Core partners in the pilot program included the Rid All Green Partnership Farm, the Garden Valley Neighborhood House (a settlement house and food pantry that also offers training programs, senior programs, tutoring, and other afterschool programs for youth), Environmental Health Watch (an environmental non-profit focused primarily on food, agriculture, and healthy homes), and Fresh By Nature Records (a Black/female owned and operated record label and music production company). The first two are located in and have been operating in Garden Valley for over 10 years, while EHW has partnered with the Rid-All Farm for several years in Garden Valley. Community events, largely conceptualized and planned by residents, used strategies of placemaking to bring community members from across Garden Valley together to share food, participate in public art projects, music performances and open mics, and to participate in several youth-oriented activities (including puppet shows, shoe-garden demonstrations, and games). In an area of Cleveland where violence and fear are significant barriers to residents' feelings of safety (there are places where residents don't feel comfortable crossing the street) placemaking activates spaces in a positive way, slowly shifting the narrative of Garden Valley, both from the outside gaze as well as in the eyes of residents.

The planning team consisted of both Garden Valley residents and outsiders working in the community, as well as founding members of the four core partner institutions. Partners worked to engage residents through a set of interrelated programs linked by the theme of "History, Health, and Healing": community events that bring residents together to push back against the isolation prevalent in many areas of food inequity. These included music, arts, and food curriculum with a focus on production and entrepreneurialism and a program designed to jumpstart and support a youth-run café with community gardeners, urban farmers, and other community businesses among its list of suppliers. Over the course of almost a year of work on the GGS pilot program, which was an initial allocation of funds in preparation for a larger 2-year grant, three community block-party events were held in Garden Valley celebrating agriculture and the culinary arts, hip-hop and music production, and community-based visual arts projects.

Methods, data collection, and project evaluation

My role in the pilot project was in project evaluation (both for internal reflection and external assessments), to leverage current funding for additional grants, and to ensure the alignment of project activities with the objectives of the grantors. I had worked with core partners and in Garden Valley for a few years prior to this grant; however, as a non-Black person with academic institutional ties not residing in Garden Valley, my standpoint was that of an outsider. My approach to this project was to rely upon the lived experiences and histories of residents for direction in evaluation, valuing the knowledge they hold about the neighborhood and its history. As a part of the evaluation and data collection process, in addition to dozens of hours of participant observation, I conducted interviews with core partners (n = 10) and facilitated in-depth discussions with the implementation team during planning meetings (n = 15). Interview and meeting questions focused on residents' visions for their neighborhood; what kind of spaces they would like to foster and live in; how community resources could be leveraged to support Black economic growth; and the role of music, food, the arts, and culture (more broadly) in residents' vision for a healthy neighborhood. This approach created a more iterative and reflective process for both evaluation and grant implementation. While the long-term sustainability of community transformation through GGS will be unmeasurable for several years to come, the purpose of this paper is to provide insight into the organizing strategies adopted within this particular community and to examine how grant money can be leveraged to catalyze structural, sustainable, and equitable change within communities.

Over the course of the pilot, the project team (including this author) conducted two separate community surveys (n = 79), convened an intergenerational planning team of Garden Valley residents and core partners, and engaged in a process of systems-based evaluation that led to the award of two additional years of funding. Planning team meetings were facilitated as a platform for decision making as well as for data collection, allowing residents and groups members to draw upon their own epistemologies of community transformation, and to establish a vision for development within Garden Valley, rather than relying on a priori standards. The pilot events, workshops, and youth-oriented curricular program (which now serves as the backbone for continued grant activities) were conceived of collaboratively. These meetings also represented an opportunity for group reflection and the establishment of an evaluation framework. Community survey instruments were developed collaboratively; they were used to evaluate events held in August and September of 2016, to solicit participation from residents interested in future projects, workshops, or activities, as well as to outline an ongoing agenda for GGS. Survey data within Garden Valley through GGS was supplemented with two other surveys from a city-led transportation project in the same neighborhood. This provided data triangulation about community needs and desires, it more than doubled our sample size, and confirmed anecdotal evidence about other challenges faced by residents in Garden Valley such as (fear of) violence and the need for more programs and opportunities for neighborhood youth. A summer curricular program, geared towards youth, was planned as a result of these surveys, and implemented during the second phase of the grant.

Both the survey instruments and planning team meetings relied on predominantly qualitative evaluation metrics to highlight the unique challenges facing a community, through deep engagement with residents to determine which neighborhood challenges are prioritized by community residents. Our hypothesis was that knowledge rooted in everyday experiences would allow us more effective and sustainable strategies to address the systemic, structural, and cross-sectional problems within Garden Valley. As Wolf et al. (2017) highlight, single-issue approaches, such as attempts to mitigate obesity through behavioral change, risk overlooking the lived experience of residents (who might *also* be struggling with low literacy or an abusive home environment). We recognized the impossibility of addressing all of the struggles in Garden Valley through creative placemaking and entrepreneurship alone. However, focusing on residents' perceptions and desires for their own neighborhood allowed for more meaningful connections between food, arts, culture, and agriculture, while also exploring various approaches to equitable community change among neighborhood residents and CD practitioners.

The collaboration between core partner institutions, residents, and people working in Garden Valley led to a majority Black planning team, with two non-Black members, including this author, out of a total of thirteen. Leadership on the project was entirely Black, reflecting the racial makeup of the community. While the project is ongoing in its second phase, it is important not only to share lessons learned during this time, but also to contribute a perspective on a different approach to CD—both its strengths and weaknesses—and how that different approach might impact community spaces.

Placemaking and entrepreneurship at the intersection of art, culture, and food

With the historical geography of Garden Valley in mind, and the collective experience of having worked in this community for several decades, the point of entry for grant activities could not be a demand for behavioral change or any prescriptive advice about what the community *should* or *could* do to transform itself. As one planning team member articulated, it could take several years to build enough trust in the neighborhood to gain the buy-in of community members, before sustainable or transformative could even begin to take place. It is from this context that the *History, Health, and Healing* approach to placemaking emerged and became essential to GGS. Within community development, most interventions are not designed to recognize the relationship between present-day inequities (including in food access) and histories and geographies of oppression (Kepkiewicz et al. 2015; Usher 2015). With this in mind, planning team members conceived of three block-party style events in Garden Valley celebrating urban agriculture, hip-hop culture, and Black culinary arts, not necessarily to highlight histories of oppression, but rather to begin healing from its impacts. These events were intentionally planned to be part of the iterative process of grant planning, and helped us to develop the successful proposal for two additional years of funding supporting youth engagement around the aforementioned themes. At pilot events, we interviewed participants, distributed surveys, and gathered more informal feedback about how Black residents in this neighborhood envisioned community transformation.

"Soul Food Saturdays" is an example of a monthly event at Rid-All that reflects the *History, Health, and Healing* focus of GGS, as well as a growing trend of vegan soul food in Cleveland. Residents experienced cuisine that retains historically and culturally important Black traditions while featuring a diversity of healthful ingredients. These events support the holistic ideology undergirding GGS, in that they expand beyond just food or agriculture, bringing community members together to talk about the challenges they are facing and discuss how they would like to see their community transform. One of the owners of Rid-All articulates this point:

> ...if we could approach this work from a food angle, we could reach a wide variety of people, everybody. [...] The activity that's generated from us being here is what we call placemaking. So it's more than just a farm; it's extended out to the broader community where families can come out and have picnics at the park, (where) women come out and walk their dogs or play with their children at the park. [...] So that becomes a residual value of a project that you can't really monetize, you can't put a dollar value on it, but the value is there because of what it does for the human experience. So there's a lot of ways of looking at how we do our work and how it plays out every day. Especially in the community. Not to mention the pride that people have in this area. This was the Forgotten Triangle: something that was basically a dumping ground (transformed into) a thriving beacon of innovation.

This trend of more "healthy" soul food in Cleveland represents the meeting point of placemaking, Black entrepreneurialism and the ideology behind *History, Health, and Healing*. Soul food originated as a culinary survival mechanism during times of slavery: a way to prepare food with inexpensive ingredients (usually high in fat and salt). Through a *History, Health, and Healing* lens, soul food—often criticized for being unhealthy—can be celebrated as an important historical tool for the survival of Black people in this country. Vegan soul food businesses within Cleveland's Black community celebrate this tradition, while modifying it to the present-day context. A practice of appreciating the present moment without anticipating or worrying about the future is inherent in southern traditions of food. *History, Health, and Healing* emphasizes and celebrates this unique history of Black Clevelanders—who are predominantly descended from enslaved ancestors, and have memories of or close relatives who were bonded through sharecropping or tenant farming—and the many ways in which community and food come together to promote health and healing. Recounting Black histories and geographies as a celebration of health and wellness instead of focusing on health disparities or less healthy food choices is a powerful form of placemaking within a community.

Engaging in holistically-minded food system work in historically marginalized and oppressed communities of color such as Garden Valley requires an historically rooted knowledge of the expansive role of food, farming, and land within the Black community: how these have been and continue to be sources of both liberation and oppression (Vernon 2015; Yakini 2015). Food is an incredibly important convener within the Garden Valley community as well as in other communities of color in Cleveland. We learned first-hand during the pilot grant that music and art play a similar role as powerful conveners of community, with a seemingly natural bridge to placemaking and community-based enterprise.

One of the founders of Fresh By Nature articulates her sentiments about music in a way that mirror how the founders of Rid-All articulate food and growing:

> …an entire economy surrounding hip-hop can grow out of communities like Garden Valley. Dance, graffiti, slam poetry, parties, albums, slams, etc., (are all a part of it). It's not so much about the music, per se, but about people working together: as long as people are working together they can elevate. Hip-hop is just the common thread.

GGS was conceived out of the belief that, while food security is essential to the well-being of any community, it will be an ineffective point of intervention if it stops with food (either self-provisioning, food assistance, or both). Partners and participants articulate a universal need for food that is mirrored by historical-cultural connections to food and land that provide important opportunities for both community building and cultural healing. And, arts and music permeate all of that: "There's no disconnect." Especially in the context of historically and spatially marginalized spaces, food system work needs to embrace a rooted understanding of the multi-faceted cultural histories embedded within communities of color.

Black growers in Cleveland not only articulate an understanding of the complexity of community needs, but also target a broad set of placemaking-related objectives in their agricultural work. Efforts are rarely focused singularly on food or the choices that neighborhood residents make around food consumption, but rather demonstrate a rootedness in place and a desire to produce space differently. In explaining the capacity of the land to promote healing, one resident touched upon some of the broader objectives of the Gardens and Green Spaces project:

> I began to look at ways to help the neighborhood heal. Because that had to be a horrific realization: you saw people going into that house and never coming back out.[1] You've lived in this neighborhood for years, maybe even decades. You observed this, even if you just moved in and moved across the street. You were a complicit part in it all. It trickled to the whole… area and beyond – so how could we bring healing to a place that's broken? The scab is laying there bare. You understand? I was blessed to be able to make a little bit of a dent in that area because (the CDC president) was kind enough to give me a parcel of land.

The ultimate goal with placemaking and among Black entrepreneurs is to help their community heal from present and historical traumas: to thrive rather than just survive. This perspective recognizes that diet-related disease, other health indicators, educational attainment, employment, and most economic, social, political, and cultural issues are both interrelated and have roots that extend beyond the social context of a particular community and the individual choices of residents. Residents participate in small enterprise as a way to create beautiful spaces, to produce urban space differently (Ekers and Loftus 2012; Torreggiani et al. 2012), to reclaim and reshape their own agricultural histories in non-oppressive ways, and, often, to disembed the production of space from the capitalist economy, re-embedding these processes in social relations (Heynen et al. 2006). GGS strives to shift the dominant negative perception of Garden Valley, which means addressing the entire socio-natural context: the physical, mental, emotional, and spiritual aspects of the community, in addition to the environmental, spatial, or health concerns generally targeted by CD initiatives.

Rooted in an historical consciousness of the trauma of marginalization experienced by the Black community, Black-led enterprises are central to the visions and theories of change held by many Garden Valley residents for the future of their community, and contributed heavily to the vision residents expressed for GGS. The resident-driven neighborhood engagement philosophy recognizes that supporting the particularities of a neighborhood's histories and geographies is central to successful transformation, including in entrepreneurialism and creative placemaking. Black entrepreneurship is embedded in the ideologies and practices of several of the core partner institutions of GGS. The entrepreneurial aspects of the grant project emerge from the particular knowledge and experience of the core partners, and are designed as sustainability mechanisms to eventually render grant funding more peripheral to the project. Thus conceived, the project depends on market demand for products and services in addition to non-profit and philanthropic funding, the latter of which has replaced a significant proportion of state funding in low-income communities over the last several decades (Ghose and Pettygrove 2014; Mitchell 2001).

During the pilot year, the planning team and core partners were constantly reminded why a more holistic approach to community development is more difficult and less common than more singularly focused programs. The complexity of confronting historical trauma within a community is something that takes time, patience, and dedication, but is also necessary if sustainable transformation is going to occur.

[1] This quote to the 2009 discovery of the missing bodies of eleven Black women on the property of Anthony Sowell, whose death sentence is currently on hold.

Year after year, Garden Valley residents see grants programs come and go without any long-term commitment. Building trust takes time, and allowing the process to be iterative and responsive to the needs of community residents takes flexibility and a willingness to not always be as successful—in the eyes of the grantor—as we might otherwise be. This work challenges dominant paradigms of community development in Cleveland not only by avoiding a top-down, expert-driven approach often funded through the city and CDCs, but by demonstrating to funders what an iterative process of community change looks like, as identified in our evolving study, work, and evaluation. In this way, we work not only to shift the power dynamic in community development, but also for more equal power relations between grantor and grantee.

The ultimate vision of Gardens and Green Spaces was to promote a transformative production of space in Garden Valley, as led by the vision of neighborhood residents, especially youth. With the award of a two-year grant, we have continued to focus on "History, Health, and Healing", and community development that supports the visions of complex and multifaceted people. Our goal is to produce Black space differently in Garden Valley; to engage in creative placemaking that lifts up the artistic, cultural, and culinary histories of this place. Our goal is to change the way that these Black spaces are perceived and experienced, for neighborhood residents and for the broader Cleveland community.

Discussion

The literature on both placemaking and entrepreneurialism highlights tensions between enacting a particular vision for community through grass-roots and community-level activities, and enabling the retreat of an increasingly austere state, thereby reinforcing ideologies of neoliberalism, individualism, and personal responsibility. Similar tensions are also communicated by Black entrepreneurs, residents in Garden Valley, and Black urban growers more broadly. Participants express a desire to participate in the capitalist economy, but also to push back against it or establish something apart from it. Several participants have expressed a desire to create spaces that would not need to be so heavily policed, indeed indicating a desire for a retreat of (at least part of) the state apparatus (Purcell 2008). In conversations and interviews with growers and entrepreneurs, the desire to form or join cooperative economic structures is frequently articulated, as is a vision for a "Black economy" that exists separately from the dominant white economy. Rather than replicating all of the tenets of "white capitalism", as the case study shows, these Black subjects envision an economic structure built explicitly to support community growth, resilience, and empowerment. According to one participant in GGS, individual success and community-oriented economies can happen simultaneously. As one interviewee described, they "both feed off each other. While (entrepreneurship) might seem individualistic and boot strapping, it's also feeding the community" (Personal Interview 2017).

In a similar vein, creative placemaking, while sometimes described in the literature as a means of validating the withdrawal of the state or neighborhood divestment, becomes a potentially radical avenue for residents to produce (alternative) spaces. In other words, residents are working outside of the confines of a state or developmentalist perspective on what (the production of) space should look like or entail. This enables them to assert their own way of being and living in the world, allowing a vision for Black space to emerge. Whether the residents of Garden Valley are able to negotiate and navigate the tensions present here depends largely on the capacity to create an economically sustainable program that continues to bring in revenue, contribute to positive community transformation, and provide economic opportunities for residents. The two-year implementation grant is now in its second year and is focused largely on how to transition from being mostly grant-funded to becoming a self-sustaining project that can remain true to the vision, needs, and desires of Black residents in creating the places and spaces that will promote healing and support a community to thrive.

Conclusion

Heynen et al. (2006) argue that the struggle for liberation will be played out in urban societies, including Cleveland. This case study of the Gardens and Green Spaces project demonstrates how this struggle can be repositioned through creative placemaking and a community-oriented entrepreneurialism. From interviews, community meetings, and other interactions with residents, it is evident that this approach touches down in very real ways for many Black Clevelanders. Many facets of life that are seldom emphasized in low-income neighborhoods (such as arts, and culture—including food culture) are not only recognized, but celebrated as integral to the fabric of a community. Black food culture has played a key role in movements organizing for Black rights, including the Civil Rights Movement, within the Black Panther Party, and in organizing for Black voter registration; this lends resonance to the approach of GGS, through an historical connection to culture, food, and the production of Black space. Black family histories of slavery, sharecropping, tenant farming, and/or participation in the Great Migration have transported culinary traditions and alternative urban imaginaries into the city. During one of the community events, a participant commented, "See how we can make space from nothing into something? We create what we want and need in our neighborhoods...we can create something beautiful in a place where there wasn't anything like that." These ideologies supporting GGS are rooted not only within the Garden Valley community, but in a collective Black consciousness and memory across the city.

The tacit goals of community development often rely upon economic development: establishing new businesses, bringing jobs to a particular area, or increasing foot traffic through a neighborhood to help businesses grow. While economic development is embedded within the GGS project, it does not emerge as an objective isolated from the goal of building resilient and supported communities. Rather, entrepreneurialism, economic growth, and creative placemaking are conceived of as a means to a more broadly imagined liberation. Resident-driven transformation is holistic, cross-sectional, and rooted in place, because the historically embedded inequities and oppressions facing the Black community are so deeply entrenched that economic empowerment alone cannot eliminate oppression or produce equitable and thriving Black spaces.

Acknowledgements The author would like to thank Kimberly Foreman, Executive Director of Environmental Health Watch in Cleveland for being a great teacher and collaborator, Tess Pendergrast for her assistance as an editor, and two anonymous reviewers for their comments on earlier versions of this paper.

References

Allen, E. 2010. Growing Community Food Systems. In *The Post Carbon Reader: Managing the 21st Century's Sustainability Crises*, ed. R. Heinberg and D. Lerch. Santa Rosa, California. www.postcarbonreader.com.

Angelo, H. 2017. From the City Lens toward Urbanisation as a Way of Seeing: Country/City Binaries on an Urbanising Planet. *Urban Studies* 54 (1): 158–178.

Bain, A.L., and F. Landau. 2017. Artists, Temporality, and the Governance of Collaborative Place-Making. *Urban Affairs Review*.

Brahinsky, R. 2013. Race and the Making of Southeast San Francisco: Towards a Theory of Race-Class. *Antipode* 46 (5): 1258–1276.

Brenner, N., and N. Theodore. 2002. Cities and the Geographies of 'Actually Existing Neoliberalism'. *Antipode* 34 (3): 348–379.

Cavanaugh, E., et al. 2014. Changes in Food and Beverage Environments After an Urban Corner Store Intervention. *Preventive Medicine* 65: 7–12.

Cummins, S., E. Flint, and S.A. Matthews. 2014. New neighborhood grocery store increased awareness of food access but did not alter dietary habits or obesity. *Health Affairs* 33 (2): 283–291.

Derickson, K.D. 2014. The Racial Politics of Neoliberal Regulation in Post-Katrina Mississippi. *Annals of the Association of American Geographers* 104 (4): 889–902.

Ekers, M., and A. Loftus. 2012. Revitalizing the Production of Nature Thesis: A Gramscian Turn? *Progress in Human Geography* 37 (2): 234–252.

Ghose, R., and M. Pettygrove. 2014. Urban Community Gardens as Spaces of Citizenship. *Antipode* 46 (4): 1092–1112.

Grewal, S.S., and P.S. Grewal. 2012. Can Cities Become Self-Reliant in Food? *Cities* 29 (1): 1–11.

Grodach, C., and A. Loukaitou-Sideris. 2007. Cultural Development Strategies and Urban Revitalization: A Survey of US Cities. *International Journal of Cultural Policy* 13 (4): 349–370.

Harvey, D. 1996. *Justice, Nature and the Geography of Difference*. Cambridge: Blackwell Publishing.

Harvey, D. 2003. Debates and Developments: The Right to the City. *International Journal of Urban and Regional Research* 27: 939–941.

Heynen, N., M. Kaika, and E. Swyngedouw (eds.). 2006. *In the Nature of Cities: Urban Political Ecology and the Politics of Urban Metabolism*. London: Routledge.

Heynen, Nik, Hilda E. Kurtz, and Amy Trauger. 2012. Food Justice, Hunger and the City. *Geography Compass* 6 (5): 304–311.

Holt-Giménez, E., and Y. Wang. 2012. Reform or Transformation? The Pivotal Role of Food Justice in the U.S. Food Movement. *Race/Ethnicity: Multidisciplinary Global Contexts* 5 (1): 83–102.

Jenkins, W.D. 2001. Before Downtown: Cleveland, Ohio and Urban Renewal, 1949-1958. *Journal of Urban History* 27 (4): 471–496.

Jessop, B. 2000. The Crisis of the National Spatio-temporal Fix and the Tendential Ecological Dominance of Globalizing Capitalism. *International Journal of Urban and Regional Research* 24 (2): 323–360.

Kaufman, J., Bailkey, M. 2000. *Farming Inside Cities : Entrepreneurial Urban Agriculture in the United States*. Lincoln Institute of Land Policy: Cambridge, Massachusetts.

Kay, A. 2000. *Art and Community Development: The Role the Arts Have in Regenerating. Community Development Journal* 35 (4): 414–424.

Kepkiewicz, L., M. Chrobok, M. Whetung, M. Cahuas, J. Gill, S. Walker, and S. Wakefield. 2015. Beyond Inclusion: Toward an Anti-colonial Food Justice Praxis. *Journal of Agriculture Food Systems and Community Development* 5 (4): 99–104.

Lawson, L. 2004. The Planner in the Garden: A Historical View into the Relationship between Planning and Community Gardens. *Journal of Planning History* 3 (2): 151–176.

Lefebvre, H. 1991. *The Production of Space* (trans: Donald Nicholson-Smith). Oxford: Blackwell.

Lefebvre, H. 1996. *Writings on Cities* (trans: Eleonore Kofman and Elizabeth Lebas). Oxford: Blackwell.

Lefebvre, H., N. Brenner, and S. Elden (eds.). 2009. *State, Space, World: Selected Essays*. Minneapolis: University of Minnesota Press.

Ley, D. 2003. Artists, Aestheticisation and the Field of Gentrification. *Urban Studies* 40 (12): 2527–2544.

Lindemann, Justine. 2019. Black Urban Grower and the Land Question in Cleveland, OH: Externalities of Gentrification. In *Back to the City: Food and Gentrification in North America*. New York: New York University Press.

Massey, D.S., and M.J. Fischer. 2000. How Segregation Concentrates Poverty. *Ethnic and Racial Studies* 23 (4): 670–691.

Michney, T.M. 2011. White Civic Visions Versus Black Suburban Aspirations: Cleveland's Garden Valley Urban Renewal Project. *Journal of Planning History* 10 (4): 282–309.

Mitchell, K. 2001. Transnationalism, Neo-liberalism, and the Rise of the Shadow State? *Economy and Society* 30 (2): 165–189.

Mitchell, T.W. 2005. Destabilizing the Normalization of Rural Black Land Loss: A Critical Role for Legal Empricism. *Wisconsin Law Review* 2005: 557–616.

Montgomery, A. 2016. Reappearance of the Public: Placemaking, Minoritization and Resistance in Detroit. *International Journal of Urban and Regional Research* 40 (4): 1–24.

Morales, A. 2009. Public Markets as Community Development Tools. *Journal of Planning Education and Research* 28 (4): 426–440.

Németh, J., and J. Langhorst. 2014. Rethinking Urban Transformation: Temporary Uses for Vacant Land. *Cities* 40: 143–150.

O'Brien, D., E. Staley, S. Uchima, and E. Thompson. 2004. *The Charitable Food Assistance System: The Sector's Role in Ending Hunger in America*. Washington, D.C.: The Congressional Hunger Center and Feeding America.

Ohio Department of Transportation (OHDOT). Cleveland Opportunity Corridor. http://www.dot.state.oh.us/projects/ClevelandUrbanCoreProjects/OpportunityCorridor/Pages/default.aspx. Accessed 10 July 2017.

Pennick, E.J. 1990. Land Ownership and Black Economic Development. *The Black Scholar* 21 (1): 43–46.

Porter, C.M. 2018. What Gardens Grow Outcomes from Home and Community Gardens Supported by Community-based Food Justice Organizations. *Journal of Agriculture Food Systems and Community Development* 8 (Suppl 1): 187–205.

Porter, M. 1995. The Competitive Advantage of the Inner City. *Harvard Business Review* 73 (3): 55–71.

Pritchett, W.E. 2003. The 'Public Menace' of Blight: Urban Renewal and the Private Uses of Eminent Domain. *Yale Law & Policy Review* 21 (1): 1–52.

Purcell, M. 2008. *Recapturing Democracy: Neoliberalization and the Struggle for Alternative Urban Futures*. New York and London: Routledge.

Purcell, R. 2009. Images for Change: Community Development, Community Arts and Photography. *Community Development Journal* 44 (1): 111–122.

Reese, Ashanté M. 2018. We Will Not Perish; We're Going to Keep Flourishing': Race, Food Access, and Geographies of Self-Reliance. *Antipode* 50 (2): 407–424.

Richmond, M. 2017. In One Glenville Neighborhood, Residents See Looming Gentrification. *WVIZ Ideastream*. http://wviz.ideastream.org/news/in-one-glenville-neighborhood-residents-see-looming-gentrification. Accessed 13 July 2017.

Safransky, S. 2017. Rethinking Land Struggle in the Postindustrial City. *Antipode* 49 (4): 1079–1100.

Saldivar-Tanaka, L., and M.E. Krasny. 2004. Culturing Community Development, Neighborhood Open Space, and Civic Agriculture: The Case of Latino Community Gardens in New York City. *Agriculture and Human Values* 21 (4): 399–412.

Sassen, S. 2000. *Cities in a World Economy*, 2nd ed. Thousand Oaks: Pine Forge Press.

Sbicca, J. 2012. Growing Food Justice by Planting an Anti-Oppression Foundation: Opportunities and Obstacles for a Budding Social Movement. *Agriculture and Human Values* 29 (4): 455–466.

Scott, J.C. 2010. *The Art of Not Being Governed: An Anarchist History of Upland Southeast Asia*. New Haven: Yale University Press.

Shannon, J. 2014. Food Deserts: Governing Obesity in the Neoliberal City. *Progress in Human Geography* 38 (2): 248–266.

Smith, N. 2008. *Uneven Development: Nature, Capital, and the Production of Space*. Athens: The University of Georgia Press.

Sommers P, Smit J. 1994. *Promoting Urban Agriculture: A Strategy Framework for Planners in North America*. Cities Feeding People Series, IRDC. Ottowa, Canada.

Stokols, D. 1995. Translating Social Ecological Theory into Guidelines for Community Health Promotion. *American Journal of Health Promotion* 10 (4): 282–298.

Sundin, E. 2011. Entrepreneurship and Social and Community Care. *Journal of Enterprising Communities: People and Places in the Global Economy* 5 (3): 212–222.

Swyngedouw, E. 1996. The City as a Hybrid: On Nature, Society and Cyborg Urbanization. *Capitalism Nature Socialism* 7 (2): 65–85.

Tornaghi, C. 2017. Urban Agriculture in the Food-Disabling City: (Re) Defining Urban Food Justice. *Reimagining a Politics of Empowerment. Antipode* 49 (3): 781–801.

Torreggiani, D., E. Dall'Ara, and P. Tassinari. 2012. The Urban Nature of Agriculture: Bidirectional Trends between City and Countryside. *Cities* 29 (6): 412–416.

Tuan, Y. 1977. *Space and Place: The Perspective of Experience*. Minneapolis: University of Minnesota.

Usher, K.M. 2015. Valuing all Knowledges Through an Expanded Definition of Access. *Journal of Agriculture Food Systems and Community Development* 5 (4): 109–114.

Vernon, R.V. 2015. A Native Perspective: Food is More Than Consumption. *Journal of Agriculture Food Systems and Community Development* 5 (4): 137–142.

While, A., A.E.G. Jonas, and D. Gibbs. 2004. The Environment and the Entrepreneurial City: Searching for the Urban 'Sustainability Fix' in Manchester and Leeds. *International Journal of Urban and Regional Research* 28 (3): 549–569.

Wilson, D. 2007. *Cities and Race: America's New Black Ghetto*. London: Routledge.

Wolff, T., M. Minkler, S.M. Wolfe, B. Berkowitz, L. Bowen, F.D. Butterfoss, B.D. Christens, V.T. Francisco, A.T. Himmelman, and K.S. Lee. 2017. "Collaborating for Equity and Justice: Moving beyond Collective Impact. Nonprofit Quarterly (Winter), pp. 1–19.

Yakini, M. 2015. What Ferguson Means for the Food Justice Movement. *Food Justice Voices: Issue 1*. https://whyhunger.org/k2-category-publications/item/2987-what-ferguson-means-for-the-food-justice-movement. Accessed 25 Feb 2016.

Justine Lindemann is a PhD candidate in Development Sociology at Cornell University. Her research explores the intersection of race, food, and urban planning and policy in Cleveland, OH. Much of the work she does is in collaboration with Cleveland residents and community organizers, striving for more innovative and inclusive approaches to land use and community development.

Agriculture and Human Values (2019) 36:879–889
https://doi.org/10.1007/s10460-019-09973-8

SYMPOSIUM/SPECIAL ISSUE

Participatory plant breeding and social change in the Midwestern United States: perspectives from the Seed to Kitchen Collaborative

G. K. Healy[1] · J. C. Dawson[1]

Accepted: 10 July 2019 / Published online: 20 July 2019

Abstract

There is a strong need to connect agricultural research to social movements and community-based food system reform efforts. Participatory research methods are a powerful tool, increasingly used to give voice to communities overlooked by academia or marginalized in the broader food system. Plant breeding, as a field of research and practice, is uniquely well-suited to participatory project designs, since the basic process of observing and selecting plants for desirable traits is accessible to participants without formal plant breeding training. The challenge for plant breeders engaged in participatory research is to consider not only how their work incorporates farmer input in developing new varieties, but also how it interacts with broader questions of food sovereignty, food justice, diversity and democratization in the food system. This article examines these issues in the context of the Seed to Kitchen Collaborative, a participatory variety evaluation and breeding project at the University of Wisconsin-Madison.

Keywords Participatory research · Seed sovereignty · Organic agriculture · Plant breeding

Introduction

Participatory research allows researchers to engage directly with broader social movements for a more equitable food system. Participatory methods ideally give researchers and practitioners an equal voice in determining priorities and long-term goals. Participatory research also includes practitioners such as farmers in the day-to-day project decisions. Participatory plant breeding (PPB), as a form of participatory research, prioritizes farmer and stakeholder input in variety development. It is the most established avenue for plant breeders to engage in local and organic food movements as well as food and seed sovereignty initiatives. However, of the four major discursive approaches to the food movement—organic advocacy, food sovereignty, community food security, and food justice (Alkon 2013), it has proven easier for PPB projects in the U.S. to promote organic advocacy and seed sovereignty, and more challenging to meaningfully engage with food sovereignty per se and food justice initiatives.

We explore the potential for PPB to engage with community-based efforts through the lens of our own project, the Seed to Kitchen Collaborative (SKC) in Madison, Wisconsin. This participatory variety evaluation and selection program engages with organic and local food movements, as well as seed and food sovereignty movements in the upper Midwest. The SKC has been successful in terms of its stated goal of promoting collaborations between plant breeders, farmers and chefs to improve selection for flavor and direct market quality. These goals easily dovetail with needs expressed by farmers and other actors in local and organic food movements. Partnerships with seed sovereignty and food sovereignty organizations have been slower to develop, despite substantial alignment of interests and long-term goals of participants. Concrete engagement with food justice organizations also remains elusive. In this article we explore some of the opportunities and challenges that the SKC has encountered. As many of these are common to participatory programs originating within academic institutions, we hope that using the SKC as a case study can provide a point of discussion for researchers navigating similar issues within their own programs.

This article is structured in three sections. The first is an overview of the SKC as it exists now. The second section discusses areas where the SKC has made tangible

✉ J. C. Dawson
dawson@hort.wisc.edu

[1] Department of Horticulture, University of Wisconsin-Madison, Madison, USA

contributions. The third discusses aspects of seed sovereignty, food sovereignty and food justice where establishing partnerships for specific participatory projects has been more challenging, even when long-term goals are shared. Historical context is provided and discussed where it illuminates the current opportunities and challenges. It is beyond the scope of this article to provide a full review of food and seed sovereignty movements or participatory plant breeding initiatives. Review articles of these topics are cited in the relevant sections.

The Seed to Kitchen Collaborative

The Seed to Kitchen Collaborative at the University of Wisconsin, Madison was initiated in 2013, in response to local farmer requests for more focused research to improve certain crops for organic production and direct marketing in the upper Midwest. The project started after a conference at the Stone Barns Center for Food and Agriculture in New York that brought together chefs and plant breeders to discus breeding for flavor and for local food systems (Gordinier 2013). In 2014, a group of five chefs, five farmers and seven UW Madison plant breeders participated in a variety trialing project, with the goal of evaluating varieties and breeding lines of vegetables for important sensory and productivity traits. At the time of writing, the project has expanded to include a core group of ten Madison chefs with additional chefs participating occasionally in Madison and other towns, over 70 farmers and gardeners from the upper Midwest, and plant breeders from universities, private seed companies and independent entities across the country. Farmers are recruited through multiple regional farmer list-serves, as well as presentations at winter conferences. Chefs are referred to the project through word of mouth and through outreach conducted through the Madison-area chefs collaborative.

Priority setting

Farmers, chefs, seed companies and other local food advocates are closely involved in planning and implementation. Each January, participants meet to share their priorities and ideas for the coming season with researchers, who also provide a summary of results from the previous season. This input guides project design for the coming year, and establishes a culture of open dialogue within the group that endures throughout the growing season. Chefs and restaurant groups provide a lucrative local market for produce, and are often a barometer for food trends that may influence customers' demand in the season to come. From an internal evaluation survey conducted in early 2018, 68% of participating farmers have adopted a variety they first saw through the SKC trials. In a few instances, farmers have begun growing a specific variety that chefs favor, leading to new income opportunities for the grower. For example, after evaluating the flavor of hot peppers for three seasons, Chef Jonny Hunter of the Underground Food Collective contracted with a local grower to produce over 1000 lb his favorite pepper, Aji Rico from Ball-Pan American Seeds, bred in collaboration with UW Madison plant breeder Jim Nienhuis, in the 2018 season.

SKC solicits submissions of varieties and breeding lines from seed companies and plant breeders around the country who serve Midwestern farmers, but who lack capacity to conduct trials in the region. Currently, vegetables are the focus, but recent tastings have included bread wheat and cider apples. The vegetables in the 2017 trials included: tomatoes, peppers, potatoes, carrots, kale, winter squash, melons, onions and specialty greens. Trials prioritize new varieties that have been developed in organic or low-input systems and for good flavor and high quality for local markets, but have not been trialed extensively in the Midwest.

Implementation

The trials employ a mother-baby participatory research design (Snapp 1999, 2002), which features a central trialing "hub", and satellite on-farm trials focused on a subset of species and varieties. The "hub" sites are the West Madison Agricultural Research Station in Southern WI and the Spooner Agricultural Research Station in Northern WI, where all trial crops and varieties are grown in organic systems, in replicated field plots. For each crop, a "check" variety is identified to help standardize results across environments. The check is always a well-known, reliable variety that provides a good benchmark for quality and productivity.

Farmers choose how many crops, and which crops and market classes within crops to trial (for example, butternut varieties of winter squash, or corno di toro varieties of pepper). This allows farmers to integrate the trials into their existing production. On-farm trials are managed according to the farmer's preferred techniques, to ensure that results are relevant. The only standardizations required are that trial varieties are not placed at the edges of fields or ends of rows, to reduce environmental variation due to edge effects, and that the check variety is replicated. Most farmers participating are either certified organic or using organic/low-input management practices.

Participating farmers receive training in conducting trials at winter meetings, and receive an instructional packet and follow-up phone calls throughout the season. As crops develop and mature, farmers keep notes including whether they would grow the variety again, how marketable it was, and their evaluation of flavor, productivity, disease resistance, and earliness. The information collected and datasheet were developed with the group of participating farmers and breeders, based on what information they would like to know about a new variety before they would consider growing it on a larger scale and what feedback the breeders wanted from farmers. Research station trials include more quantitative data such as yield, dates of maturity, and disease scores. The research crew also tastes all varieties after receiving a short training, and this tasting data is used to narrow down the list of varieties for participating chefs.

Chefs evaluate varieties for flavor and culinary quality at monthly flavor evaluations from June through January. Each event is hosted at a different restaurant, and the chefs represent a variety of cuisines, as well as dining services for two large institutions. One or two tasting events per year are also organized for chefs outside of Madison. The flavor evaluation methods vary slightly by crop, but all evaluations employ rapid sensory analysis methodology designed to make quantitative flavor evaluation accessible to end-users, rather than requiring a panel of trained professionals, which can be expensive and impractical (Dawson and Healy 2018).

The on-station, on-farm and culinary trials provide complementary types of information on variety performance. This information is compiled at the end of the season for breeders and seed companies to use in improving varieties for organic and direct-market production systems in the upper Midwest and that other farmers can use to make decisions about what to grow. When chefs find a variety they really like, they can work with farmers to produce it, providing a ready market for that variety and creating demand for seed of varieties developed for organic and local food systems.

Outreach

The project reaches the general public through an event called Farm to Flavor where SKC participating chefs prepare dishes featuring favorite varieties. Community members are invited to enjoy the chefs' creations and learn from plant breeders, seed companies, and farmers about the varieties that perform best in local conditions. Every season, the SKC also participates in a number of public events, to engage gardeners and other vegetable enthusiasts in evaluating quality. In Madison, the SKC has partnered with the Centro Hispaño to do a hot pepper tasting at their annual community festival, and has done variety tastings with Alice's Garden in Milwaukee. The SKC has also facilitated chef tastings of indigenous squash varieties and wild rice with the Intertribal Agricultural Council. The SKC has participated in several events with Madison-area non-profits focused on food access and equity, giving away transplants or running a vegetable tasting booth. Each year, the SKC does taste tests at the West Madison Agricultural Research Station Urban Horticulture Open House and the Spooner Agricultural Research Station Twilight Tour with a few hundred people attending in each location. These collaborations not only allow the SKC to incorporate public perspectives on culinary quality into flavor evaluations, but also to build relationships with gardeners and farmers who may be interested in collaborating on seed saving or variety development projects in the future.

Seed to Kitchen success: connecting plant breeding to social movements for food system change

Plant breeding for organic systems and regional agriculture

The rapid growth of SKC is in part because it fills a need in the growing organic and local food and seed sectors. There has been double-digit growth in the US market for organic produce over the last 20 years, and local food sales increased from $1 billion in 2005 to $6.1 billion in 2012 (Low et al. 2015). Consumers expect superior flavor, texture, appearance, culinary quality and nutritional value from produce grown organically, especially when purchased fresh from direct market operations (Bonti-Ankomah and Yiridoe 2006). Farmers selling direct to consumers therefore prioritize flavor when choosing varieties, but flavor may not be a priority in plant breeding programs targeting wholesale markets. In addition, an estimated 95% of varieties grown on organic farms are not bred for organic environments, though research has shown that varieties developed in conventional systems often under-perform in organic systems and that varieties intended for organic production benefit from direct selection in organic systems (Hoagland et al. 2015; Lammerts van Bueren and Myers 2011; Murphy et al. 2007; Reid et al. 2011; Renaud et al. 2014).

This means that organic farmers may lack access to the genetics best suited to their production systems and market demands. Even when organic priorities align with conventional priorities (in the case of yield, for example) the same plant may express the desired trait differently in organic versus conventional production systems (Lammerts van Bueren and Myers 2011). For example, a plant bred to be highly productive in a high-input environment may yield poorly when planted in an organic environment without soluble fertilizers with immediately available nutrients. Likewise, a plant bred to excel in organic systems may not be the top performer in conventional trials. Furthermore, many breeding programs at global seed companies focus on broad adaptation across a wide range of environments rather than focused adaptation to a specific region. This reflects a long-term trend toward plant breeding for larger conventional seed markets, rather than a decentralized system of breeding programs able to prioritize important traits for regional adaptation. (Dawson et al. 2008; Annicchiarico et al. 2012; Hoagland et al. 2015).

There is natural complementarity between breeding for organic systems, and breeding for a region of intended use. Since organic farmers cannot use synthetic inputs to standardize their environments, local biotic and abiotic conditions (such as pests and soil quality) heavily influence organic farm ecology. There is therefore a need for crops adapted to regional organic conditions. As the organic seed sector grows, many plant breeders at public institutions and small to mid-scale seed companies are selecting and evaluating varieties in organic environments within a region, with local farmers, to ensure that their production and market needs are reflected in breeding priorities (Chiffoleau and Desclaux 2006). Produce quality and flavor are often high priorities for breeders interested in supporting organic systems and direct-market growers.

The number of plant breeders working on organic seed systems is still relatively small compared to the growth in organic food markets. Not all regions of the US have plant breeders working on all crops of importance. Trialing varieties bred in regions with some similarities in environmental conditions can identify varieties that may have top performance in multiple regions. For example, many varieties selected in New England have done well in the SKC trials in Wisconsin. However, this translation from one environment to another cannot be assumed; some varieties that excel in their region of selection perform poorly in the SKC trials. Small to mid sized seed companies and public institutions often do not have capacity to do multi-regional tests, so regional programs like SKC are necessary to allow farmers in the upper Midwest to provide feedback to breeders in other regions. This type of trialing also lays the groundwork for direct farmer involvement in selection, using the best varieties to create breeding populations for selection within the region. In doing so, SKC trials increase farmer engagement in determining what trait combinations will be available in the upper Midwest.

Participatory research in organic and local food movements of the Global North

SKC prioritizes aligning project goals with farmers' needs, engaging farmers in variety evaluation, and returning trial results to breeders whose varieties will be available to Midwestern growers. In doing so, SKC employs a model of participatory research akin to Participatory Plant Breeding (PPB), though most SKC farmers are not directly engaged in performing selection or developing breeding populations. In using PPB as a model, SKC has joined a growing number of breeding and variety evaluation programs in North America and Europe adapting PPB methodology, which originated in the Global South, to a Northern context. Decentralized participatory research supports organic variety development for environments characterized by complex biotic and abiotic stress, rather than forcing farmers to rely on results from research environments that may not be representative of organic systems (Bonneuil and Demeulenaere 2007; Campanelli et al. 2015; Wolfe et al. 2008; Dawson et al. 2008, 2011).

PPB is distinguished from formal centralized breeding strategies by its complimentary priorities of breeding for adaptation to a specific region or production environment, and engaging farmers in the variety development process. Formal breeding programs often centralize selection (often in a highly controlled and optimal environment, usually a research station), and prioritize uniformity, high yield, and broad adaptation across a wide geographical range. Conversely, PPB prioritizes decentralized selection (on farms in the environment of intended use), genetic diversity, adaptation to marginal environments, and farmer involvement (Ceccarelli 1997). PPB has roots in researcher realizations that farmers in marginal growing environments who could not irrigate and fertilize crops based on research recommendations were not using varieties developed by formal plant breeding programs (Evenson and Gollin 2003). PPB arose to "reverse the historical trend of separation between farmers and plant breeders", meet the agronomic needs of marginalized farmers, and bring farmers' knowledge into decisions made about variety improvement (Cleveland and Soleri 2002).

PPB methodology has proven to be highly adaptable. Increasingly plant breeders working on organic systems use participatory methodologies to meet the particular environmental and quality needs of organic farmers. The initial motivation for PPB was in part due to extreme differences in the environments of research stations and farmers' fields in the Global South (Ceccarelli 1997). In contrast, most formal breeding programs in North America and Europe conduct selection under conditions closely resembling conventional farmers' fields. These selection environments, however, rarely resemble organic farms because of the difference between organic and conventional environments and because organic farms are quite diverse in their management and markets. Participatory research methods are therefore all the more important in organic systems. In France, for example, the "Farm Seed Network" (Réseau Semences Paysannes) helps farmers develop wheat varieties that yield well on their particular farms and meet cultural expectations for bread flour (Chable et al. 2014; Dawson et al. 2011). In 2014 plant breeders from the University of Wisconsin-Madison released an open-pollinated sweet corn variety bred in partnership with farmer Martin Diffley, and High Mowing Organic Seed Company (Shelton 2014). These programs help organic farmers gain greater control of their variety choices and seed production, and can increase variety diversity (Sperling et al. 2001).

Another difference between PPB in the Global North and the original PPB programs of the Global South is the type of crops typically bred using participatory methods. While many PPB projects in the Global South focus on staple crops, many of the projects currently underway in the US engage specialty crop farmers who grow dozens of crops and multiple varieties of each crop. For vegetable crops, seed production is often more technically demanding and time consuming than for grains, and many farmers do not save seed. Because diversified vegetable farms operate intensively during the summer season, few farmers have the time to maintain a breeding population on-farm. PPB methodologies must be adapted to allow farmer participation at different phases of the breeding process. Allowing farmers to participate based on their time and availability ensures that outcomes reflect their needs, even if their schedules cannot accommodate active on-farm selection in early-generation breeding populations (Witcombe et al. 2005; Dawson and Goldberger 2008; Dawson et al. 2008).

Incorporating participatory trials into breeding programs has proven to be an effective tool for taking advantage of farmers' expertise without requiring the time commitment of on-farm selection. This has been found to result in superior outcomes when breeding for organic or low-input conditions (Dawson et al. 2011; Ceccarelli et al. 1992; Ceccarelli 1994, 1996; Soleri et al. 2002). It is critical, however, for programs to do more than simply implement on-farm trials. Farmers must be involved in setting goals and in evaluating trials with information returned to breeders. SKC has worked closely with farmers to design trial protocols to accommodate their schedules and to focus on the information most useful to farmers and breeders. This means that while most farmers involved in the SKC trials do not conduct on-farm selection, they are full participants in evaluation and impact breeding program goals.

Current PPB programs in the US also diverge from earlier PPB programs in engagement with non-farmer stakeholders. In traditional farming communities, production and consumption are often tightly linked. In the US there are many key players in local food systems populating the supply chain between producers and consumers, and consumers are increasingly engaged in making decisions about what their local food systems should look like. In response to this complex food system structure, a growing number of PPB projects involve chefs, bakers, brewers, produce buyers and community supported agriculture (CSA) customers in evaluating varieties and determining breeding priorities. The SKC has been involving chefs in evaluation over five seasons, including selection in early generation breeding populations with a high level of variability. The resulting varieties reflect the tastes and opinions of the chefs as well as of farmers with on-farm trials. This level of feedback from multiple food system stakeholders is an asset to plant breeders working to develop varieties adapted to organic farms in the upper Midwest.

Breeding for organic systems is important to farmers and consumers. But the organic movement is often charged with elitism, since organic food is often more expensive than conventionally grown food and marketed to privileged communities. It is important for plant breeders to consider how their work intersects other paradigms of food system change such as food sovereignty, and food justice. The SKC project is now starting on-farm selection with several farmers, with the intent of building farmers' capacity to develop their own varieties. In moving from on-farm trialing to on-farm selection, the SKC intersects more directly with seed sovereignty initiatives in the upper Midwest, which seek to go beyond incorporating farmers' perspectives in selection to returning some ownership of the seed supply to growers.

Up to this point, however, the impacts of the SKC have been primarily within the formal (market-based) seed system, and projects involving community-based seed organizations have been slower to develop. One fundamental challenge for plant breeders is that our current seed system, including the organic sector, exists in an agricultural landscape that is almost entirely based on market solutions and individual (rather than community-based) decision-making. The lack of institutional or legal recognition of community-based rights, ownership and governance is echoed in struggles faced by the larger food sovereignty movement. It can be seen in the apparent contradictions of local activism for autonomy from the global food and agricultural systems, and national or global activism to change policy in existing governance structures to institutionalize food sovereignty and community based initiatives (Shattuck et al. 2015).

Though the SKC attracts participants from throughout the local food system, the group struggles to create programming that transcends the organic and local movements and contributes to food sovereignty and food justice. These movements challenge the assumption that adjusting people's purchasing habits to include more local or organic food, or more locally adapted and organic seed, will yield an ecologically sustainable and socially just food system. Rather, collective action to decentralize power in the food and seed system is needed for farmers and communities to gain the skills, knowledge, and seeds needed for thriving local food systems. The larger food sovereignty movement also struggles with the tension between building local alternatives outside of the globalized agricultural system and efforts to make the current system more fair (Shattuck et al. 2015), so this issue is not specific to organic agriculture or seed systems.

Seed to Kitchen Opportunities: plant breeding, seed sovereignty and food justice

Community seed sovereignty

Advocates for seed system democratization in the US and Europe often look toward leaders in the Global South for successful models of food and seed sovereignty such as Vandana Shiva and Navdanya, the organization of over 600,000 farmers she founded, and the global Via Campesina, which originated in grassroots food sovereignty initiatives across Latin America (Altieri and Toledo 2011). The Via Campesina movement defines food sovereignty as "the right to healthy and culturally appropriate foods, produced through ecologically sound and sustainable methods, and [peoples'] right to define their own food and agriculture systems" (2009). This right to control over food and agriculture systems includes a necessary element of "seed sovereignty" which Vandana Shiva defines as the "sovereign rights of farming communities to their seed and plant genetic resources" (Kloppenburg 2014; Shiva 1997). These movements are generally independent of academic or professional plant breeders and often actively reject the trends of privatization of germplasm that both private and public sector plant breeding programs adopted in the twentieth century.

In the Global North there are a growing number of organizations and communities working on seed sovereignty outside of the commercial seed realm. Indigenous communities across the U.S. and Canada have created or revived traditional seed saving programs, such as the Tesuque Pueblo Seed Bank in New Mexico (Brascoupe 2002) and the Native American Seed Sanctuary, a collaboration between the St. Regis Mohawk/Akwesasne tribe and Seedshed, a local nonprofit (Seedshed 2017). Hundreds of local seed libraries now exist throughout the U.S., as communities explore and redefine the concepts of seed and food sovereignty for themselves (Greene 2013).

Retaining control over a seed supply means that communities retain the ability to perpetually adapt food and agriculture systems to meet shifting environmental, economic and cultural realities. Without some measure of control over their seed system, farmers must confront the unique demands of their particular environmental and cultural context with a set of genetic tools (commercial seeds) that were not created for their particular situation. This circumstance not only makes farmers unnecessarily vulnerable, it also reduces the agricultural biodiversity that results from having millions of farmers around the world continuously selecting the plants they steward.

Though seed laws are somewhat different in Europe than in North America, both have created an environment in which spaces for informal seed exchange and farmers' right to save and replant seed are increasingly limited, favoring individual and corporate ownership and market-based seed systems (Kloppenburg 2010). Many authors have addressed issues related to privatization of genetic resources and the effects of intellectual property rights (IPR) on farmers' rights to select and save seed. The following section focuses specifically on how IPR regimes affect the intersection between seed sovereignty movements and participatory plant breeding projects, and how SKC navigates the tension between different frameworks for seed ownership and control.

Intellectual property rights

In the 1970s, the landscape for intellectual property protection in the US moved away from only relying on biological forms of protection such as hybrid seed to also include legal forms such as Plant Variety Protection (PVP) certificates. PVP certificates still allow farmers to save seed for their own use and breeders to use protected varieties for new crosses. However, in recent years more companies are turning to utility patents for new varieties, which do not include an exemption for breeders to use a variety in breeding or for farmers to save seed. Many public-sector breeders do not support recent trends towards more restrictive forms of ownership, however, university plant breeders are typically required to turn over rights to any new varieties to their university technology transfer office, which decides on pursuing IPR and licensing (Tracy 2017). This creates a problem if varieties are created by a collaborative effort. Naming all participants as co-inventors is one way to partially address the issue, but does not guarantee that the university will protect and license the variety in a way that allows those involved in its creation to access the variety or produce their own seed if they choose.

While many plant breeders at universities have good relationships with their technology transfer offices, most feel that the increasingly legalistic environment restricts their ability to exchange seed with other programs, even those at other universities (Shelton and Tracy 2017). A few initiatives have arisen to respond to this situation. One, primarily led by public sector plant breeders, aims to establish a set of best practices for exchanging germplasm and releasing varieties from university programs that would maintain farmers' rights to save seed and breeders' ability to continue to use publically developed varieties in breeding (Dawson et al. 2018). Another, the Open Source Seed Initiative (OSSI) promotes a "protected commons" for seed, in which plant breeders can release, and farmers may access, varieties under a pledge that no restrictive protection will be sought for a given variety or any of its progeny (Kloppenburg 2014). This does not preclude the generation of revenue from varieties, and many breeders who have contributed varieties to OSSI either sell seed of these varieties themselves or receive royalties from a seed company selling seed of their varieties.

It is important to acknowledge that there is some tension between the concepts of food sovereignty and the open source concept. Some communities may decide that sharing culturally and economically important seed in any commons, even a protected one, may not be in their best interest. For example, the Réseau Semences Paysannes (Peasant Seed Network) in France places their work on seed sovereignty in the framework of the collective rights of a community, to be maintained for the benefit of the community. This contradicts the position of genetic resource collections and public sector institutions, which tend to promote agrobiodiversity as a public good (Demeulenaere 2014). In this case, a participatory breeding program exists in collaboration with the French National Agricultural Research Institute (INRA), but selection and control of the seed stays with the farmers' organization (Dawson et al. 2011). Many indigenous seed sovereignty movements steward seed as a living part of their community, and do not want it appropriated by external entities. This is in keeping with principles articulated by Indigenous communities that understand food as sacred, part of the relationship with the natural world that defines culture and community (Desmarais and Wittman 2014). Food sovereignty and seed sovereignty are inextricably linked.

Though an increasing number of public, private and independent plant breeders advocate for a more open exchange of germplasm, they still need to be compensated for their work in order to continue their plant breeding programs. Breeders interested in alternative seed ownership models (such as OSSI) often struggle to balance the need to earn income from breeding work with the desire to make varieties and breeding lines open and available for anyone to adapt to their own production systems. In addition, there are differing opinions on how open the exchange of germplasm should be. The public sector best practices initiative maintains that it is appropriate to restrict the commercial sale of seed of a particular variety but not to prohibit continued use of the variety in breeding, so that it stays a common genetic resource. The OSSI initiative maintains that there should be no restrictions at all on commercial use of varieties as long as downstream recipients also pledge not to restrict use. Community seed sovereignty initiatives often reject individual ownership of varieties but support community ownership and restrict use of those varieties outside the community. These are all valid points of view in different contexts, and ensuring that the wishes of communities or breeders are respected is an important component of plant breeders' engagement in seed sovereignty initiatives.

In the SKC, for example, encouragement for farmers to take up their own variety selection and breeding work exists in tandem with a firm commitment to upholding the wishes of the variety originators. This includes variety protection options that allow compensation for plant breeders, as long as these protections allow for the continued use of the variety in breeding and on-farm seed saving. In the case of community ownership of traditional varieties, it would be inappropriate to incorporate these varieties into a program that might later fall under the ownership of a university. However, plant breeders could serve as resources for community led initiatives if requested, building a project separate from the university-led project that specifically responds to community priorities.

Potential joint projects of this type require researchers to acknowledge that the relationship between public institutions in the US and indigenous peoples and communities of color has been fraught with appropriation and paternalism (Hassel 2004; Caldwell et al. 2005). It takes time to repair trust. Even if researchers were not directly involved in this history, the institution they work for likely was. Researchers need to understand the history of their institution and not dismiss community concerns as outdated, as many instances of historical appropriation are still producing inequalities today. Trust can be built, however, often around specific shared projects. One of the plant breeders participating in the SKC works with a local indigenous community to help preserve a variety they have been growing for generations. While some tribal leaders approached the plant breeder requesting assistance, other influential members of the tribe were opposed to working with the university. A willingness on the part of both the researcher and the community to engage in conversations led to the eventual success of the partnership. The plant breeder describes that process as one of building trust, starting small and growing a relationship that truly reflects the desires of the community (personal communication 2017).

By working with the Intertribal Agricultural Council, the SKC has contributed in a small way to chef's engagement with native foods. For example the SKC coordinated a wild rice tasting with multiple strains of wild rice, and a tasting of indigenous squash varieties without either crop being part of university-run trials. There is a need to take time to discover if and how a university partnership could benefit organizations involved in seed sovereignty initiatives. A history of exploitation by more privileged groups leads to reticence to contribute seeds of important food crops to trials or breeding programs. Even with the best intentions, plant breeders need to accept that not all seed should be shared, and that not all variety stories are theirs to tell. For some communities, the cultural continuity provided by retaining exclusive control of ancestral seed stocks is more important than potential markets or adding more diversity to breeding efforts outside their control.

Food justice initiatives

In Wisconsin, the farmers, breeders, eaters, and chefs engaged in organic food organizations are predominantly white. However, there is a high level of interest, creativity, expertise and commitment to food system reform coming from people of color in the state, who participate in a number of food justice organizations. Milwaukee in particular has a strong recent history of food justice activism, as the birthplace of Growing Power, Alice's Garden, Walnut Way, and the Fondy Food Center, which facilitates fresh local food consumption by connecting farmers' of color with farmers markets in the predominantly African American communities in Northwest Milwaukee. These organizations are often focused on making healthy local food more accessible to communities that have historically been marginalized and subject to structural racism in the food system. These efforts are often framed in the language of community food security, and include work to shift from emergency food provision such as food banks to employment that provides living wages, work to increase the quality of available food in neighborhoods that have often been deserted by larger scale grocery stores, and work to increase urban food production through urban farms and gardens (Heynen et al. 2012).

It is a priority for the SKC to involve farmers and eaters of all backgrounds in trialing and breeding projects. There are no restrictions on the scale of farms involved in trialing and evaluating varieties. Some participating growers operate at a garden scale, and several trial partners are urban farms, growing small trials of culturally and economically important crops and returning observations to the program. The SKC provides seed, so seed cost is not a barrier to participation. This simple act of paying attention to how a set of varieties perform in one's backyard or neighboring lot over the course of a few seasons can help return power to a community, by equipping people to advocate for and take a role in creating better seed options for their particular circumstances. However, there are substantial institutional challenges to truly meeting the goal of engaging communities of color. While the SKC includes several farmers and chefs of color, their participation is on an individual level. There is a need for plant breeders and programs like the SKC to engage with broader food justice initiatives, often led by community organizations that historically have been ignored or excluded from university agricultural programs.

The SKC works to build relationships with these organizations, in order to more concretely contribute to both seed sovereignty and food justice. It is vital to give this process time. One of the primary barriers for both universities and non-profit partners is the grant-dependent nature of most projects. It is difficult to find resources for long-term initiatives to build trust and partnership. Though funds for outreach are often available in research grants, the metrics used are usually numbers of attendees at presentations, field days and educational bulletins published. Taking the time to deeply understand community priorities, and ensure that projects are developed in service to genuinely shared goals, does not fit neatly into grant-funded outreach activities. Similarly, university faculty are primarily rewarded for research outcomes, and continued grant funding depends on demonstrating productivity through publications and traditional outreach activities. With pressure for impact metrics and short timelines for demonstrating this impact, SKC has struggled to prioritize the longer-term relationship building that is critical but takes time away from grant-funded activities.

For public plant breeders, it is easy for long-term goals to take a back seat to the more immediate goals of publications or varieties developed. For organizations focused on food justice, the connection between growing healthy, culturally appropriate food and developing well-adapted, collaboratively created varieties may be clear, but the time horizon of expected impacts may be too long to prioritize over more pressing needs. Improving varieties for local food systems does not directly impact structural racism or economic exclusion in the short term. In these cases, the best course of action may be for plant breeders to become better allies of food justice organizations, committing to listening and learning so that the priorities of food justice advocates are incorporated in long-term planning for participatory programs.

Ways forward

Truly decentralizing power in the seed system would mean urban and rural farming communities growing, selecting, saving, trading and replanting their own seed. However ideal this vision may be, achieving it in the current farm system is impractical and, for many, undesirable. Though there is growing interest in on-farm seed production, selection and breeding, most farmers do not produce their own seed. Many farmers feel the complexity of farming is great enough without adding seed production, and prefer to buy seeds from a trusted company, to ensure quality. The desire to purchase seed may contradict traditional definitions of seed sovereignty, but it is a reality that movements for food and seed system reform must accommodate. As food sovereignty activism grows in diverse countries and eco-regions, it looks different in each of these regions, shaped by local specificities of history, culture, and political systems (Shattuck et al. 2015).

In the US, the recent growth in small seed companies, particularly vegetable seed companies focused on organic systems, presents an opportunity to provide locally adapted varieties to farmers. Sovereignty in this context may come from a strong collaborative relationship with a regional seed company or public plant breeder committed to making regionally adapted varieties available to local farmers. A growing number of public plant breeders now partner with small, regionally-focused seed companies while simultaneously involving farmers in variety development. These relationships increase farmers' power over their seed supply and contribute to seed sovereignty. The SKC has created opportunities for feedback to travel between seed companies, farmers and end-users such as chefs. Each of these stakeholders' power, then, rests not only in their ability to buy seed that reflects their values but to have a voice in which varieties are available and how those variety stories are told.

For example, the SKC works closely with Oregon State University (OSU) plant breeder Jim Myers, who submitted his 'Indigo Rose' tomato to the SKC trials in 2014 and 2015. Though it was a popular variety in the Northwest, the SKC farmers were unhappy with the Indigo Rose flavor and performance in Midwestern summer conditions. The SKC provided this feedback to Myers and he sent a number of experimental submissions the following season. One of these experimental tomatoes had Indigo Rose as a parent, but it ripened earlier and the flavor interested the SKC chefs. With Dr. Myers' permission, the SKC trialed this experimental tomato with an urban farmer, Mark Voss, who markets tomatoes directly to chefs in Madison. There was still some variation in the experimental population, so Voss selected plants with superior flavor and a dark pink complexion, and members of the SKC will work with Voss and Myers on potential release of a variety for the Midwest. In this instance, Voss's participation in the seed system was more profound than a seed purchase. Even if he ends up purchasing seed of the tomato he helped develop, that variety's presence in the market amplifies his voice and elevates the priorities of his customers and fellow farmers.

The SKC is uniquely equipped to help farmers transition from simply growing trials to performing selection and improving their own varieties. This is attractive to farmers selling directly to consumers or chefs who value novelty in their produce offerings, as well as adaptation to local farm environments. As the project has gained recognition, the SKC has received requests for seed-saving education from farmers and community groups interested in performing basic selection to further adapt crops to their environments and regain some control over their seed supply. In addition to annual field days at the West Madison Agricultural Research station, the SKC has offered one seed saving and seed systems class as part of a food systems course offered by the Urban Ecology Center in Milwaukee and several workshops on conducting on-farm trials with the Organic Seed Alliance. With increased staff capacity, the SKC could do more seed-saving and on-farm breeding education, in partnership with organizations already focused on farmer education. These skills are increasingly relevant to communities trying to revive local seed stewardship mechanisms.

The community engagement element of the SKC aligns with the educational goals of a diverse cohort of graduate and undergraduate students. Since it's founding 42 undergraduate and graduate students (including 14 Hispanic students and students of color) have worked with the SKC in some capacity, contributing their own connections and cultural knowledge to the program. This has helped the leadership team build new partnerships and envision new directions beyond the original scope of the program. The SKC will continue to recruit students from diverse cultural and academic backgrounds, and the project will continue to benefit from their interest in food justice, food security and participatory research. In an environment of limited support for public plant breeding, the SKC has been able to offer students a rare academic experience that combines the science of plant breeding with the cultural and social dimensions of food.

Conclusion

SKC is now deeply connected in the Madison, WI community and throughout the state of Wisconsin. The project has successfully built connections between farmers, chefs and plant breeders to help guide the selection of vegetable varieties for local and organic agriculture in the upper Midwest. This project has expanded participation in decision-making about what seeds will be available in the future, thereby increasing seed and food sovereignty in the region, as well as supporting the local and organic food movements.

Working with many farmers throughout the Upper Midwest and having a presence in academic and non-academic educational settings has helped SKC establish relationships with diverse communities engaged in food production. These communities are interested in finding or preserving varieties uniquely suited to their production and cultural context. However, there are still challenges in making the project truly representative of communities in the upper Midwest, a challenge that exists within the broader organic agriculture movement as well. Building trust with communities who are under-represented in university led-projects is an important goal for programs like SKC seeking to link research with broader movements for food system reform.

Breeding for organic systems and regional adaptation is essential to the continued success of the organic and local food movements, but it is equally important for plant breeders to consider how their work intersects other paradigms of food system change such as food sovereignty and food justice. Failing to do this will limit plant breeders' ability to promote a truly resilient and ethical food system. As more plant breeders become interested in participatory research, better methods will emerge, and more farmers will become active in efforts to build a seed system rooted in local control and ecological growing practices. There are no easy answers to questions of ownership, privilege, and power in the food system, but creating open spaces for respectful dialogue with diverse participants will hopefully lead to more genuine partnerships between academic institutions and community organizations.

Acknowledgements The authors would like to thank the SKC participants and the seed sovereignty and food justice organizations that the SKC interacts with for their energy, dedication and hard work. This paper would not have been possible without the many discussions we have had with all of you. We would also like to thank the University of Wisconsin-Madison, the USDA North Central SARE Program and the Ceres Trust for funding that has supported the Seed to Kitchen Collaborative.

References

Alkon, A.H. 2013. Food justice, food sovereignty and the challenge of neoliberalism. In *Food sovereignty: a critical dialogue*. September 2013: 13–14.

Altieri, M.A., and V.M. Toledo. 2011. The agroecological revolution in Latin America: Rescuing nature, ensuring food sovereignty and empowering peasants. *Journal of Peasant Studies* 38 (3): 587–612.

Annicchiarico, P., L. Pecetti, and R. Torricelli. 2012. Impact of landrace germplasm, non-conventional habit and regional variety selection on forage and seed yield of organically grown lucerne in Italy. *The Journal of Agricultural Science* 150 (03): 345–355.

Bonti-Ankomah, Samuel, and Emmanuel K. Yiridoe. 2006. Organic and conventional food: A literature review of the economics of consumer perceptions and preferences. Final report submitted to the Organic Agriculture Center of Canada. www.organicagcentre.ca. Accessed 3 May 2014.

Bonneuil C., and E. Demeulenaere. 2007. Une génétique de pair à pair? L'émergence de la sélection participative, in F. Charvolin, A. Micoud et L. K. Nyhart, dir. Les sciences citoyennes. Vigilance collective et rapport entre profane et scientifique dans les sciences naturalistes. Ed. de l'Aube, 122–147.

Brascoupe, C. 2002. Rekindling the fire of traditional agriculture. *Biodiversity* 3 (3): 26.

Caldwell, Joyce Y., Jamie D. Davis, Barbara Du Bois, Holly Echo-Hawk, et al. 2005. Culturally competent research with American Indians and Alaska Natives: Findings and recommendations of the first symposium of the work group on American Indian research and program evaluation methodology. *American Indian and Alaska Native Mental Health Research* 12 (1): 1–21.

Campanelli, G., N. Acciarri, B. Campion, S. Delvecchio, F. Leteo, F. Fusari, P. Angelini, and S. Ceccarelli. 2015. Participatory tomato breeding for organic conditions in Italy. *Euphytica* 204 (1): 179–197.

Campesina, Via. 2009. Nyéléni declaration. *Journal of Peasant Studies* 36: 673–676.

Ceccarelli, S., S. Grando, and J. Hamblin. 1992. Relationship between barley grain yield measured in low-and high-yielding environments. *Euphytica* 64 (1–2): 49–58.

Ceccarelli, S. 1994. *Specific adaptation and breeding for marginal conditions. Breeding fodder crops for marginal conditions*, 101–127. Dordrecht: Springer.

Ceccarelli, S. 1996. Adaptation to low/high input cultivation. *Euphytica* 92 (1): 203–214.

Ceccarelli, S., E. Bailey, S. Grando, and R.N. Tutwile. 1997. Decentralized, participatory plant breeding: a link between formal plant breeding and small farmers. The International Center for Agricultural Research in Dry Areas (ICARDA). Aleppo, Syria.

Cleveland, D.A., and D. Soleri. 2002. Farmers, scientists and plant breeding: knowledge, practice and the possibilities for collaboration. In *Farmers, scientists and plant breeding: Integrating knowledge and practice*, ed. D.A. Cleveland and D. Soleri, 1–18. Oxon: CAB International.

Chable, V., J. Dawson, R. Bocci, and I. Goldringer. 2014. Seeds for organic agriculture: Development of participatory plant breeding and farmers' networks in France. *Organic farming, prototype for sustainable agricultures*, 383–400. Dordrecht: Springer.

Chiffoleau, Y., and D. Desclaux. 2006. Participatory plant breeding: The best way to breed for sustainable agriculture? *International Journal of Agricultural Sustainability* 4: 119–130.

Dawson, J.C., V.M. Moore, and W.F. Tracy. 2018. Establishing best practices for germplasm exchange, intellectual property rights and revenue return to sustain public variety development. *Crop Science* 58: 469–471.

Dawson, J.C., and G.K. Healy. 2018. *Flavor evaluation for plant breeders. Plant Breeding Reviews*. New York: Wiley.

Dawson, J.C., and J.R. Goldberger. 2008. Assessing farmer interest in participatory plant breeding: Who wants to work with farmers? *Renewable Agriculture and Food Systems* 23 (3): 177–187. https://doi.org/10.1017/S1742170507002141.

Dawson, J.C., K.M. Murphy, and S.S. Jones. 2008. Decentralized selection and participatory approaches in plant breeding for low-input systems. *Euphytica* 160 (2): 143–154.

Dawson, J.C., P. Rivière, J.-F. Berthellot, F. Mercier, P.D. Kochko, N. Galic, S. Pin, E. Serpolay, M. Thomas, S. Giuliano, and I. Goldringer. 2011. Collaborative plant breeding for organic agricultural systems in developed countries. *Sustainability* 3 (8): 1206–1223.

Demeulenaere, E. 2014. A political ontology of seeds. The transformative frictions of a farmers' movement in Europe. *Focaal: Journal of Global and Historical Anthropology* 69 (2014): 45–61. https://doi.org/10.3167/fcl.2014.690104.

Desmarais, A.A., and H. Wittman. 2014. Farmers, foodies and first nations: Getting to food sovereignty in Canada. *The Journal of Peasant Studies* 41: 1153–1173.

Evenson, R.E., and D. Gollin. 2003. Assessing the impact of the Green Revolution, 1960 to 2000. *Science* 300 (5620): 758–762.

Gordinier, J. 2013. Sowing a change in kitchens. The New York Times. http://www.nytimes.com/2013/09/25/dining/sowing-a-change-in-kitchens.html?_r=2. Accessed 12 January 2014.

Greene, K. 2013. Seed library movement from roots to bloom. Hudson Valley Seed Company. http://hudsonvalleyseed.com/blog/the-seed-library-movement-from-roots-to-bloom/. Accessed 6 November 2017.

Hassel, C. 2004. Can diversity extend to ways of knowing? Engaging cross-cultural paradigms. *Journal of Extension* 42: 2.

Heynen, N., H.E. Kurtz, and A. Trauger. 2012. Food justice, hunger and the city. *Geography Compass* 6: 304–311.

Hoagland, L., J. Navazio, J. Zystro, I. Kaplan, J.G. Vargas, and K. Gibson. 2015. Key traits and promising germplasm for an organic participatory tomato breeding program in the US Midwest. *HortScience* 50 (9): 1301–1308.

Kloppenburg, J. 2010. Impeding dispossession, enabling repossession: Biological open source and the recovery of seed sovereignty. *Journal of Agrarian Change* 10: 367–388.

Kloppenburg, J. 2014. Re-purposing the master's tools: The open source seed initiative and the struggle for seed sovereignty. *Journal of Peasant Studies* 41 (6): 1225–1246.

Lammerts van Bueren, E.T., and J.R. Myers (eds.). 2011. *Organic crop breeding*. New York: Wiley.

Low, S.A, A. Adalja, E. Beaulieu, N. Key, S. Martinez, A. Melton, A. Perez, K. Ralston, H. Stewart, S. Suttles, S. Vogel, and B. Jablonski. 2015. Trends in U.S. local and regional food systems: A report to congress. USDA- ERS. https://www.ers.usda.gov/amber-waves/2015/januaryfebruary/the-size-and-scope-of-locally-marketed-food-production/. Accessed 6 November 2017.

Murphy, K.M., K.G. Campbell, S.R. Lyon, and S.S. Jones. 2007. Evidence of varietal adaptation to organic farming systems. *Field Crops Research* 102 (3): 172–177.

Open Source Seed Initiative. 2017. OSSI 'freed' seed. http://osseeds.org/seeds/. Accessed 6 November 2017.

Reid, T.A., R.C. Yang, D.F. Salmon, A. Navabi, and D. Spaner. 2011. Realized gain from selection for spring wheat yield are different in conventional and organically managed systems. *Euphytica* 177: 253–266.

Renaud, E.N.C., E.T. Lammerts, M.J. van Bueren, F.A. van Paulo, J.A.Juvik Eeuwijk, M.G. Hutton, and J.R. Myers. 2014. Broccoli variety performance under organic and conventional management systems and implications for crop improvement. *Crop Science* 54: 1539–1554.

Seedshed. 2017. Projects: Native American seed sanctuary. https://seedshed.org/native-american-seed-sanctuary/. Accessed 6 November 2017.

Shattuck, A., C.M. Schiavoni, and Z. Vangelder. 2015. Translating the politics of food sovereignty: Digging into contradictions, uncovering new dimensions. *Globalizations* 12: 421–433.

Shelton, A.C. 2014. Plant breeding for organic agriculture in the United States: A new paradigm. Doctoral dissertation, The University of Wisconsin-Madison.

Shelton, A.C., and W.F. Tracy. 2017. Variety development in the US public sector. *Crop Science* 57: 1823–1835. https://doi.org/10.2135/cropsci2016.11.0961.

Shiva, V. 1997. *Biopiracy: the plunder of nature and knowledge*. Boston, MA: South End Press.

Snapp, S.S. 2002. Quantifying farmer evaluation of technologies: The mother and baby trial design. In *Quantitative analysis of data from participatory methods in plant breeding*, ed. M. Bellon and J. Reeves. New York: Springer.

Snapp, S.S. 1999. Mother and baby trials: A novel trial design being tried out in Malawi. *Target Newsletter of the Southern Africa Soil Fertility Network* 17: 8.

Soleri, D., D.A. Cleveland, S.E., Smith, S. Ceccarelli, S. Grando, R.B. Rana, D. Rijal, and H.R. Labrada. 2002. Understanding farmers' knowledge as the basis for collaboration with plant breeders: methodological development and examples from ongoing research in Mexico, Syria, Cuba and Nepal. *Farmers, scientists and plant breeding: Integrating knowledge and practice*. Cleveland and Daniela Soleri. CAB International: 19–60.

Sperling, L., J.A. Ashby, M.E. Smith, et al. 2001. A framework for analyzing participatory plant breeding approaches and results. *Euphytica* 122: 439–450.

Tracy, W. 2017. Keeping Public Plant Cultivar Development in the Public Interest. In *Proceedings of the Summit on Intellectual Property Rights and Public Plant Breeding*, ed. W.F Tracy, J.C Dawson, V.M Moore, and J. Fisch. . August 12–13, 2016 Raleigh NC. www.agronomy.wisc.edu/ipr-summit/.

Witcombe, J.R., S. Gyawali, et al. 2005. Participatory plant breeding is better described as highly client-oriented plant breeding. II. Optional farmer collaboration in the segregating generations. *Experimental Agriculture* 42: 79–90.

Wolfe, M.S., J.P. Baresel, D. Desclaux, et al. 2008. *Euphytica* 163: 323. https://doi.org/10.1007/s10681-008-9690-9.

G. K. Healy is a Research, Education Associate for Organic Seed Alliance in the Midwest region and an Outreach Specialist with UW Madison. She holds a joint Masters degree in Agroecology and Organic Horticulture from UW-Madison, where she studied with Dr. Julie Dawson. Kitt's masters research focused on conducting tomato variety trials for short-season organic production, and engaging chefs and local farmers in participatory breeding and evaluation projects.

J. C. Dawson is an Associate Professor in the Department of Horticulture at the University of Wisconsin-Madison. Her background is in organic plant breeding and participatory research and her research interests include organic and participatory plant breeding, crop genetic diversity, urban agriculture, and support for beginning farmers.

Agriculture and Human Values (2019) 36:891–902
https://doi.org/10.1007/s10460-019-09946-x

SYMPOSIUM/SPECIAL ISSUE

To save the bees or not to save the bees: honey bee health in the Anthropocene

Eleanor Andrews[1]

Accepted: 9 May 2019 / Published online: 20 May 2019

Abstract
As honey bee colonies continue to perish at high rates, beekeepers are divided on how best to keep bees healthy and productive. In this article, I describe the tensions between conventional beekeepers and a new wave of beekeepers hoping to "save the bees" through a more "natural" approach to beekeeping. Drawing on animal studies and multispecies literature, I show how beekeepers in both camps are constrained by the reality of the Anthropocene: novel ecologies, shifting baselines, and the hybridity of honey bees themselves—part wild animal subject to environmental change, part industrial organism, embedded in circuits of migratory pollination. Beekeepers on both sides are investing in genetics as a solution to honey bee health problems. Thinking with bees helps deepen the literature on multispecies encounters and interrogate the idea of sustainability in agriculture, while thinking with the Anthropocene prompts us to ask what "saving the bees" even means in today's changing world.

Keywords Beekeeping · Honey bees · Anthropocene · Novel ecologies · Hybridity

Abbreviation
CCD Colony collapse disorder

Introduction

Every year for the last ten years, over a quarter of the honey bee colonies in the United States have died (Bee Informed Partnership 2018). They are facing deaths by a thousand cuts, including pests, disease, and changed landscapes. Beekeepers are split about how best to respond, and their divide illuminates some of the dilemmas of agriculture and animal husbandry in the contemporary era of the Anthropocene. Drawing from scholarship on multispecies and animal geographies, I analyze the tensions between conventional beekeepers and a new wave of beekeepers hoping to "save the bees" with a more "natural" approach to beekeeping. These different approaches are both constrained by the reality of the Anthropocene: novel ecologies, shifting baselines, and the hybridity of honey bees themselves—part wild animal and environmental indicator, part industrial organism, embedded in circuits of migratory pollination. Changing beekeeping practices are part of a broader discussion about whether honey bees are wild or domesticated, and what their hybridity means for production and conservation goals (Hovorka 2017; Weis 2015). In short, thinking with the Anthropocene troubles what it means to "save the bees" and whether they even need saving at all.

The Anthropocene

In 2000, the "Anthropocene" was proposed as a name for the contemporary era, in recognition that that the magnitude of human activities is now so large as to be visible in Earth's geological records, with changes in atmospheric chemistry, ocean acidification, global declines in biodiversity, and more (Crutzen and Stoermer 2000). The term quickly entered popular discourse and scholarly literature, as a shorthand for the transformation for the planet. A key dimension of these changes are "novel ecologies" which, by turns, have been described as "rambunctious" gardens that we humans have made but do not fully control, "monsters" that we have made and to whom we owe our care, and "sites of struggle" in political economy (Marris 2011; Latour 2011; Smith 1996; Robbins and Moore 2013, p. 12). The Anthropocene has also drawn together the physical and social sciences,

✉ Eleanor Andrews
esa57@cornell.edu

[1] Cornell University, Ithaca, NY, USA

challenging that stubborn and vexing nature-society divide (Lorimer 2012). Other sorts of binary thinking, black-and-white ethical reasoning, and outmoded categorizations have been similarly scrambled (e.g., species characterized as weedy or welcome, wild or domestic, invasive or native, etc.) (Lorimer 2012).

Despite—or perhaps because of—its popular and scholarly adoption, the term is widely contested. Does it signal resignation or responsibility? What does it mean to accept human domination of the environment (Crist 2013)? What does it tell us about the place and power of humans? It seems paradoxically to make humans seem all-powerful on the one hand, but powerless on the other, different from other animals, but ultimately one of them ourselves, subject like any other to changes in the environment around us. Does the narrative of the Anthropocene justify questionable interventions, particularly those that entrench economic and environmental policies favoring the few over the many (Crist 2013; Johnson et al. 2014)? And does it effectively collapse responsibility for global environmental change, pinning the blame on an undifferentiated "humanity" instead of the smaller subset of humans largely responsible for planet-wide changes (Malm and Hornborg 2014)? Does it distract from the political economy of environmental change, and would it be better called the Capitalocene (Haraway 2015)? At worst, perhaps, it does all of these things. But at best, it might be understood as a conceptual upgrade that helps us better understand the empirical reality of environmental change. Perhaps it can be harnessed as a crisis or a moment of productive reckoning, as it shows divergent paths, new new spaces for creative and democratic experimentation, and new forms of agency in a "permanently polluted world" (Johnson et al. 2014; Buck 2015; Collard et al. 2015; Liboiron et al. 2018).

And what has the Anthropocene meant for nonhuman life? Other species are going extinct at a rapid clip in what has been called the "sixth extinction" (Kolbert 2014). Even species not considered threatened are experiencing localized declines, extinctions, and range contractions, as part of a more generalized "de-faunation" (Young et al. 2016). These missing animals have been called "ghosts" (Weis 2015; Tsing et al. 2017). But the Anthropocene is about animal life as much as animal death. Some generalist animals are more abundant than ever before; these are the ones, like rats and pigeons, that we often call pests. Humans have become a force in evolutionary biology, transforming animal bodies into industrial organisms through selection and breeding (Russell 2011; Schrepfer and Scranton 2003). Indeed, the contemporary era is marked by animals' "confinement, incarceration, discipline and subjection," exemplified by zoos, labs, and concentrated animal feeding operations (Watts 2000, p. 295, in Collard 2014). A few species of animals are therefore growing in absolute numbers, as billions of animals are raised every year as livestock or for other human uses. Tony Weis has called this "commodifaunation": the exploding numbers of some species that are ever more deeply integrated into global industrial agricultural circuits and transformed into commodities (2015). This is not new; the 20th century saw the increasingly efficient conversion of grain to meat and other animal products (Cronon 1991; Boyd 2001). But these trends continue, with the increasing production of these animals ever more hidden from sight, even as their consumption is growing and growing.

A particularly fruitful vein, therefore, of the scholarly literature on the Anthropocene is in animal studies writ large (including contributions from geography, science and technology studies, anthropology, and more), with its meditations on multispecies encounters, the posthuman and more-than-human, and new sorts of "kin" and "companion species" (Human Animal Research Network Editorial Collective 2015; Kirksey and Helmreich 2010; Haraway 2015, 2008; Lorimer et al. 2015; Collard 2014; Hovorka 2017; Weis 2015; Philo and Wilbert 2000). Scholars in this "species turn" are reconceptualizing human-nonhuman relationships and ecologies as assemblages (Comaroff 2017). Calls for inclusive relationality is key, moving away from control and a turn towards "living with" nonhumans (Haraway 2008; Lorimer and Driessen 2013). Much of this work insists on the possibilities for new kinds of care, "respectful and accountable relationships with multispecies," while being careful not to overlook or understate power relations (Collard et al. 2015, p. 323; Comaroff 2017). Scholarship on ongoing efforts to "rewild" animals and spaces has been described in this vein as a challenge to what have been called the Edenic sciences: conservation biology, restoration ecology—fields that seek to undo the damage that humans have wrought (Robbins and Moore 2013).

In using animals to "think with" (in the tradition of Lévi-Strauss), it has been discovered that insects are particularly good, because of their tangled relationships to humans—and bees, perhaps, "best of all" (Beisel et al. 2013; Nimmo 2013). Indeed, honey bees (*Apis mellifera* L.) bring into relief some of the major issues within the Anthropocene: the many facets of the human-honey bee relationship (Phillips 2014; Lezaun 2011; Moore and Kosut 2013), their species-being and our social construction of it (Kosek 2010; Tsing 1995; Nimmo 2013), the complexity and normativity of science, including its biases and blind spots (Suryanarayanan and Kleinman 2012, 2016; Maderson and Wynne-Jones 2016; Silverman 2013; Watson and Anthony Stallins 2016), to name just a few. Despite the wealth of social scientific literature on the human-honey bee relationship, I hope nonetheless to add something new to the conversation, with an especially empirical approach focused on on-the-ground debates and strategies to sustain honey bee health. Despite the many incisive critiques of the term described above, I believe the idea of the "Anthropocene" clarifies the dilemmas beekeepers face as they try, simply, to keep their bees alive.

Honey bee health challenges

Modern beekeeping is over a century old. Some challenges are perennial: cold winters in the North, queen failure, pests, diseases, and more. Success in beekeeping depends on careful observation and experience, tacit knowledge, familiarity with the local environment, and a little bit of luck. But it has become even more difficult in recent years. A parasitic mite, *Varroa destructor*, was introduced to the United States in the 1980s and has begun to vector increasingly virulent diseases. Microsporidian diseases like *Nosema* are widespread, with infection patterns changing and no effective treatment available. Pests and pathogens quickly spread, as migratory pollination operations put honey bees from all over the country in contact with one another. These direct stressors are exacerbated by a decline in the acreage of honey bee forage in the US, because of agricultural intensification and development. Furthermore, some pesticides pose a problem. Their effects are increasingly sublethal, meaning that they do not kill bees directly on contact, but there is increasing evidence they interact synergistically with one other, as well as with ingredients such as fungicides, previously considered to be safe and therefore not subject to label restrictions (Mullin et al. 2010). Indeed, *all* of these direct and indirect factors interact; for instance, diseased and poorly nourished bees are less able to withstand exposure to pesticides. Thus, the beekeeping literature, "once pastoral and instructional is now a bookshelf of books on disease treatment, migratory beekeeping, killer bees, urban beekeeping, and colony collapse disorder" (Mason 2016, p. 1).

In recent years, annual losses have averaged about 30% (Bee Informed Partnership 2018).[1] While these losses are not cumulative, as colonies can be replaced by splitting existing colonies in two, keeping bees alive from year to year has become more difficult. Indeed, there two ways of measuring the health of managed honey bees: the yearly rate of losses and the total number of colonies (Steinhauer et al. 2018). The two numbers are only somewhat related: honey bee colonies can be split into multiple colonies in a year (usually not more than two or three). So although annual losses are likely higher than they were a few decades ago, the percentage of annual losses is different from the total number of colonies. In the US, the total number of colonies has declined steadily since World War II, except for a slight rise in the last decade, a trend that will be further discussed below (National Agricultural Statistics Service 2017).

Unfortunately, much of the media and even social scientific literature has muddied these issues, conflating Colony Collapse Disorder (CCD) with broader challenges and declines (Watson and Anthony Stallins 2016; Nimmo 2015). CCD was identified in 2006 after some colonies dwindled and died in unusual circumstances, and in the years that followed, it was a source of concern for many beekeepers, possibly responsible for a double-digit percentage of total colony losses. Its incidence has since dramatically decreased, to the point that one leading apicultural scientist said in 2016 that he had not seen a case in five years (Nordhaus 2016). It is difficult to estimate the total number of colonies that perished from CCD, given the lack of colony-level data, the possibility of error or deception in self-reporting as well as the inability to verify those reports, and the inconclusive nature of the main symptoms, many of which may indicate other maladies or manifest differently in different regions (vanEngelsdorp et al. 2010). These factors are wrapped up in competing visions of what is at stake in the industry, the tension between precaution and action, and a broader political economy of epistemology, making debates over CCD contentious and fraught—just as they are for honey bee health more broadly (Kleinman and Suryanarayanan 2012; Suryanarayanan and Kleinman 2016; Watson and Anthony Stallins 2016). Relatedly, an arguably disproportionate fraction of media attention has focused on the risks of pesticides, which are an important dimension of honey bee health, but only one of several, interacting factors.

More broadly, measuring the health of a superorganism is complex, or, put more bluntly, "Defining a dead colony... is not so straightforward" (Steinhauer et al. 2018, p. 1; Silverman 2013). Identifying health problems is more difficult than in the past in part because stress and disturbance now are a regular part of the colonies' environment (Silverman 2013). Furthermore, the synergistic interaction of factors has produced a diversity of discourses around honey bee health issues: some parties are attempting to isolate singular causes, others are trying to understand multiple, interacting factors; and yet others are pointing to industrial agriculture more broadly as the culprit (Watson and Anthony Stallins 2016; Suryanarayanan and Kleinman 2016).

Beekeeping has had to keep up with these threats, becoming more "more laborious, costly, and time-consuming" (Lezaun 2011, p. 742). As one researcher described it during a bee conference, what beekeepers must do today is "not like your grandfather's beekeeping, and not like your father's beekeeping—not even like your own beekeeping" from years past. But there was already a kaleidoscope of different beekeeping practices. These are based on local environmental conditions (hence the saying among beekeepers

[1] This survey data on colonies and losses has many limitations, but it is the best available.

that "all beekeeping is local") and beekeeper motivations and preferences (hence another saying, "two beekeepers, three opinions"). More significantly, beekeepers are loosely divided into hobbyists and commercial beekeepers. While nearly all of the colonies in the US are managed by commercial beekeepers, who depend wholly on beekeeping for their livelihood, these commercial operators are vastly outnumbered by hobbyists.

These different beekeepers have taken different approaches to honey bee health challenges (Suryanarayanan and Kleinman 2016; Watson and Anthony Stallins 2016; Maderson and Wynne-Jones 2016). Scholars have described the divide between "natural" and "conventional" beekeeping, or "backwards" and "scientific" beekeeping (Moore and Kosut 2013, p. 56). ("Backwards beekeeping" is a description embraced by some practitioners.) More colorfully, one writer has called it "the Hatfield-McCoy treatment-nontreatment feud" (McNeil 2016, p. 1139).

To be clear, this divide in practice is more of a continuum; beekeepers in the real world make choices that place them along it, rather than entirely at one end or the other. Sociologists Lisa Jean Moore and Mary Kosut profile a "center left" beekeeper who uses chemical treatments sparingly, when he feels there is no other choice (2013, p. 56). But the rift has been significant and bitter, and so while Richie Nimmo has written that "CCD is exactly the sort of crisis that we should anticipate facing with increasing frequency if we are living in the Anthropocene," it is perhaps not for the reasons that he describes, namely the physical unsustainability of modern agricultural systems and our increased dependence upon them (2015, p. 195). It is instead the contentiousness of this divide, the politics and pushback around what is seen as natural or necessary that makes honey bee health problems so emblematic of the Anthropocene.

Methods

This analysis of beekeeping comes from 5 years of participant observation and interviews (2013 to 2018). During this time, I observed clubs, classes, and conferences (20 organizations, at least 35 meetings, including 20 full days of classes or statewide, regional, and national conferences). I conducted 54 semi-structured interviews with beekeepers, beekeeping educators, extension agents, and apiculture research scientists.[2] These individuals are mostly based in New York; a handful with a national or regional profile are based in other states or provinces. Their experience ranges from 0 to 40 or more years of beekeeping, and over half of them make some money from beekeeping, in a variety of ways. Several are or have been employed in the apiculture industry: government inspection programs or university extension, a beekeeping magazine, and a beekeeping supply company. I oversampled respondents that I expected to explicitly describe their approach as more "natural," for instance, by recruiting heavily from one bee club under a larger organizational umbrella called the Local Living Venture (formerly known as the Sustainable Living Project) and attending classes by beekeeping educators advertising a more natural or organic approach to beekeeping. My respondents therefore are not representative of beekeepers in the US or New York more broadly.

The conventional approach to honey bee health

The advent of modern beekeeping in the 19th century produced a network of professional beekeepers, researchers, and educators across agribusiness, government, and academia (particularly agricultural colleges). In the 21st century, in response to CCD and other honey bee health challenges, this small but active network marshaled its resources for research and outreach, leveraging existing programs in institutions across the country, including, for example, the five existing federal honey bee labs. The 2008 Farm Bill mandated research and reporting on CCD, to identify the causes, increase colony survival, and improve the overall profitability of the industry.

But even as these programs were put in motion and concern over CCD faded, colony survival did not measurably improve. The industry continued to conduct research, developing genetic stocks of bees resistant to *Varroa* mites, new formulations of feed and dietary supplements, and chemical and non-chemical pest controls. With new technologies, researchers began to identify ever more pathogens at the molecular level, and integrate environmental and in-hive data from smart sensors. New regulations were passed to address the global trade and domestic transport of honey bees and the use of pesticides and antibiotics. Extension programming grew to include a range of education, outreach, and business support for commercial beekeepers, including disseminating best management practices through tech transfer teams and online resources, and Master Beekeeping programs. In 2015, a federal-level Pollinator Health Task Force provided further funding for research and directed federal agencies and state governments to create plans to protect pollinators, including honey bees. While the plans have been contentious (largely around questions of stakeholder representation), they have attempted to clarify best management practices, and provided funding for habitat enhancement as well as research, outreach, and education more broadly. This approach to honey bee health is not a dramatic overhaul, but has supported beekeeper livelihoods

[2] Informed consent was obtained from all individual participants included in the study.

and a profitable industry more widely through a continued reliance on research and a focus on the management of mites and other pests.

A more "natural" approach?

While the beekeeping industry was working to produce new tools and techniques to manage honey bee health, mainstream media picked up on CCD and the other challenges. A broad interest in honey bee health was launched, attracting a new wave of beekeepers across the US. By all accounts, club memberships around the country grew rapidly, with a new wave of beekeepers very different from the old guard. Many of these new beekeepers were out to "save the bees" and thus approached beekeeping differently, diverging from conventional beekeeping both practically and philosophically.

One beekeeper I interviewed contended that "the most common starting point for novices is a tacit rejection of 'traditional beekeeping,'" including fruits of the sort of research described in the previous section. Indeed, a recent article that describes "bee-centered beekeeping" notes that there is "considerable disagreement about what these methods should and do entail," simply that it stands "in stark contrast to large-scale, commercial beekeeping" (Ellis 2018, p. 13). Much new information in books, blogs, and classes is structured as a critique of the conventional approach and industrial agriculture more broadly, skeptical of the system that "keeps beehives moving in the short term while contributing to their overall demise in the long run" (Suryanarayanan 2014, p. 266). And while most adherents to this approach are new, the critics also include the former president of the American Beekeeping Federation, a lifelong commercial migratory beekeeper with 12,000+colonies, and at least one well-established apicultural scientist and author of a classic textbook on honey bees. His recently published manifesto calls for a rethinking of beekeeping practices wherein "every idea goes against conventional wisdom" (Winston 2015).

To these ends, many beekeepers are trying to pioneer innovative approaches, experimenting with what is variously called "treatment-free," "apicentric," or "natural" beekeeping. Treatment-free beekeepers are wary of chemical treatments for *Varroa* mite infestations. They may use essential oils or probiotics, even organic acids, but not synthetic miticides, seeing the cure (for mites) as worse than the disease, and pointing to studies that show that most of the pesticides in hives are in fact the ones that beekeepers apply themselves, to manage mites and other problems (Mullin et al. 2010). Indeed, one of my contacts, who occasionally attended meetings of his local club, had been dissuaded by keeping bees himself because of the "diseases and mite, insect, and mammal pests that pose challenges [and the] toxic materials that are represented as being likely needed to manage [them]." Many treatment-free beekeepers also try to approximate the conditions that feral honey bees survive in, with different hive styles, less equipment (e.g., plastic or wax guides for bees to build comb on), and fewer hive manipulations.

Framed generously, this wave of beekeepers is a band of plucky new freethinkers helping to save the bees by moving away from an overreliance on industrial agriculture, trying out unconventional methods for sustainable stewardship, approaching beekeeping holistically and with care, and rejecting easy solutions like synthetic chemicals and dietary supplements. But many conventional beekeepers are frustrated with this narrative, and claim that these new, treatment-free beekeepers are in fact contributing to the very problem that they claim to want to solve. First, they say, this approach results in elevated honey bee mortality. The principles of evolutionary selection behind treatment-free beekeeping are predicated on the thesis that only the strong survive, which, by extension, means the weak perish. One interviewee suggested that "if the Humane Society knew what was going on, [it] would probably have a word or two to say" implying that treatment-free beekeepers are letting their colonies suffer. This is because while treatment-free beekeeping may in many ways be a simpler approach, that simplicity is deceptive, and it is more difficult in practice to keep colonies alive. Managing disease with fewer tools requires a deep understanding of the intricacy of colony dynamics, such that beekeepers must be "better biologists than almost everybody" (according to one interviewee). Or, as one of my interviewees colorfully complained, "if it was as simple as sticking your bees in a top bar hive [a common alternative and more "natural" style of hive], do you not think that we would have [expletive] done that? Do you really think that... commercial beekeepers literally don't care enough to do something that [expletive] easy?!" Finally, even if adherents to treatment-free beekeeping are willing to tolerate higher losses in the short term, for the sake of longer-term shifts in beekeeping practices and agriculture, their neighbors might not be. Untreated bees can spread parasites and disease if they become "mite bombs": colonies that, collapsing under mites or other pressures, may be abandoned by their bees or visited by bees that rob the honey from the dying colony—both of which can spread hitchhiking *Varroa* mites or other diseases. These bees are thought to be responsible for late summer spikes in neighboring hives' mite populations, making those colonies more susceptible to collapse. Infestations can spread as many as four miles in just a few months, given how far bees fly (McNeil 2016). Some even suggest that annual losses may have in fact risen in recent years precisely because of the new wave of beekeepers. Thus noninterventionism comes instead to look more like irresponsibility.

So, while some alternative approaches may seem promising, they tend to avoid a serious consideration of the demands of commercial beekeeping and idealize the ability of colonies to fend for themselves. One interviewee lamented how the new wave of beekeepers were doing a disservice to the industry by framing commercial beekeepers "forced by circumstance" to use strong chemical miticides as "uneducated or cruel or stupid or counterproductive to the welfare of bees as a whole." Indeed, this commercial/hobbyist difference is key, even though it is not the only axis of difference between conventional and treatment-free beekeepers. Hobbyist beekeepers can take a more experimental or laissez-faire approach, since their livelihoods do not depend on the health of their bees. They can lose bees and money, absorbing the costs of failure in the service of a larger endeavor. But this very freedom and low stakes may also simply lower the engagement and responsibility of the beekeepers; several interviewees said they had seen lots of people start in the last 5-10 years but only a few who had stuck with it. The one to two thousand commercial beekeepers in the US, by contrast, have invested some or all of their livelihoods in beekeeping, and there is evidence that they lose a smaller fraction of their bees every year (McNeil 2016, p. 1141). So, at worst, the new beekeepers look well-meaning, but also clueless and even dangerous as they ignore a hard-won scientific consensus about what the main honey bee health problems are, minimize the expertise and investment of individuals with decades of experience, let their own bees suffer, spread disease and parasites, and quit after a few years of failure.

A brief episode from one club exemplifies the divide: a novice beekeeper posted on the list-serv to ask for help with an unconventional hive style and mentorship from an "organic" beekeeper. The primary authority within the club wrote back with the provocation that "many of the people promoting natural beekeeping [know] very little about beekeeping," comparing them to beekeepers from the 1700s who believed fanciful myths. He continued, "there is not one right way to do things… However, there is an infinite number of wrong ways." Another member chimed in, writing: "We all want to become NATURAL beekeepers, but without Bees we are Not Beekeepers!" (meaning that beekeepers who adhere to such methods exclusively are risking their colonies). Another member of the club suggested that his more conventional friend (the initial authority figure) was acting as a "voice for the chemical industry" in echoing the sorts of recommendations made by task forces stuffed with agribusiness representatives. Indeed, conventional beekeepers have not always had the last word. One of my interviewees gently mocked a commercial beekeeper in his club who had boasted that he was going to "show amateurs how to keep bees" by overwintering some colonies in upstate New York (rather than bringing them south for the season, as is typical for commercial operators)—but then lost them all.

Finally, it is important to remember that this divide is really a continuum. So while the distinction is analytically useful in some ways, in the end, to classify real beekeepers as one or the other oversimplifies complex approaches. Beekeepers themselves do use the categories—but usually to define and disparage *other* people's approaches. Most of my interviewees did not fall neatly into one category or another, particularly those with more than just 1 or 2 years of experience. They were conflicted about whether to use synthetic mite treatments: some had changed their practices over time; others did one thing and taught another. Several more conventional beekeepers did not treat chemically for mites ("I haven't treated my bees ever. I've never put poison in my hives.") and one formerly treatment-free beekeeper, after having lost all of his colonies several times, has begun to try some (organic) treatments, as well as feeding his bees when forage is scarce, and educating himself more. Interviewees across the spectrum spoke about both individual management challenges and broader environmental stressors, even though conventional/scientific beekeepers tended to frame bee health as an issue at the level of the individual operation, and natural/backwards beekeepers tended to frame it as a symptom of broader environmental degradation. One successful part-timer spoke about both in the same breath: "We want to try to get rid of many of those terrible pesticides, but I think most of the dying, most of the collapse of hives, are due to poor management."

Apiculture in the Anthropocene

What follows is an analysis of how beekeepers on both sides of this divide articulated dilemmas around management choices for honey bee health. At the time of the interviews, I was aiming to understand divergent approaches to so-called "sustainable beekeeping." As I began to analyze the interviews, however, it became clear that many of my respondents were describing the limitations of their management choices in the "post-wild world" of the Anthropocene (Marris 2011).

Novel ecologies and shifting baselines

Beekeepers with several decades of experience have seen firsthand how the environment has changed (Maderson and Wynne-Jones 2016). The bloom calendar has shifted to the extent that, in the northeastern US, some flowers bloom on average nearly two weeks earlier than they once did. One respondent from a beekeeping family complained that "my grandparents kept a pretty good record of when the flows started… [but the old calendar] doesn't work anymore, because the environment has changed." Agricultural practices increasingly limit the amount of bee forage available around the edges of fields and roadsides. Pesticides contaminate the landscape, and although the new formulations rarely kill bees outright, they do interact with each other, accumulate in wax, and weaken bees against other maladies, plus residues of past generations of pesticides can remain in the soil and wax for decades. New pests and pathogens have arrived (*Varroa* and tracheal mites in the 1980s) while others have changed (e.g., a new form of *Nosema* disease has different symptoms than the old). In the 1980s, a new subspecies of Africanized honey bees was introduced, and these "killer bees" now outcompete European honey bees in parts of the southern US.

On the plus side, some new (invasive) plant species have turned out to be good honey plants (e.g., purple loosestrife and Japanese knotweed), making honey bees more reliant on disturbed landscapes, particularly edge areas and along roads (vanEngelsdorp and Meixner 2010). Indeed, mowing less and expanding these weedy strips is touted as a win–win solution for pollinator health and saving on municipal upkeep. This is true in urban areas as well: one beekeeper in New York City described a "smorgasbord" of "infinite nectar" for his bees, as landscapers across the city—and particularly in the nearby conservatory garden—regularly rip out plants that have stopped flowering and replace them with new ones in bloom during the growing season.

How does this shape beekeeping today? One new beekeeper spoke about the framework that she uses to situate herself among different approaches, one that expressly took into account today's changed environment:

> You want to let [the bees] do their own thing. But is that even possible anymore in our climate and in this environment?... I think that we should do things that are most environmentally sustainable, rather than "natural"... it's like the vaccine debate... I don't want to use antibiotics and prescription medicine for every little thing that I do, and I want to live a healthy life, where I can prevent most ailments, but at the end of the day, I need vaccines for certain things, because it's just impossible in our current world to prevent myself from getting things that you need to get vaccinated for. Why would it be any different with farming? ...That's just how our world is, whether we want to accept it or not.

Interviewees repeated this over and over: "how we deal with our bees cannot be the same way that I was taught to deal with my bees even 15, 20 years ago." When I asked one beekeeper and apicultural scientist about whether "sustainable" was a useful concept, she replied "If we didn't have mites, yes. The problem is we have mites." Another explained that "a lot of the practices that... the newer beekeepers want to participate in are really very much reminiscent of the way we used to keep bees, before there were lots of pest and agricultural problems." He was sympathetic to these practices but did not think they were possible without "massive changes in... agriculture." Another beekeeper told me that the "old model doesn't work anymore, just like you can't farm like Dad used to." Beekeepers must account for these new species and relationships or their bees will likely die.

A few participants described the limits and counterproductivity of trying to control the environment. One, a biology professor and hobbyist beekeeper, described how reducing the number of insect species in a field leads to "huge imbalance problems" where the remaining species "become pests themselves." She explained that when "you try to limit or control [the bees'] environment in some way, you're having massive effects in subtle ways that you can't even appreciate until it's right in your face." Other beekeepers in New York State have begun lobbying against efforts by the state government to control the invasive species that their bees make a substantial crop of honey from.

And these "novel ecologies" of the 21st century are only the latest change (Davis et al. 2011). Honey bees themselves were brought from the Old World in the 17th century (although there are thousands of species of native bees).[3] One respondent said that beekeepers were responsible for honey bee health "because we brought them over here," presumably meaning North America. Others described how beekeeping itself was "unnatural," making a more "natural" approach to it nonsensical: "There is a push where we don't do anything with our bees, we don't treat them, we just leave them, basically, in a natural environment. But I think as soon as you put them in a box, that's unnatural. So, my take is that we need to do everything we can for them." In the same way, one considered a treatment-free approach an "oxymoron": "You're putting them in a box, so that's not free of any treatment at all." In other words, in beekeeping at all, beekeepers are engaging in some sort of implicit contract, as they are in relationships to other livestock and pets (Weis 2013).

Finally, the changed environment has made scientific research more difficult, given the impossibility of maintaining control colonies and isolating experimental variables. One bee researcher presenting to a bee club meeting lamented several times that "we could actually do real experiments back then," a decade or two ago, before certain viruses and pesticides were so widespread. Another scientist had determined through previous work for the Environmental Protection Agency that nowhere in the US is completely free of pesticides. Now, when conducting any research on

[3] Honey bees may vector diseases that affect native bees and compete with them for resources more broadly, although the ecological evidence is mixed on the overall effect of honey bees on native bees.

bees, he tests even his control colonies for pesticide residues in the wax, pollen, and bees' bodies, which might interfere with experimental conditions. In sum, many respondents expressed the need for beekeepers to take care of bees as they are, and where they are, not as they would like them to be.

Hybridity

Bees as "ghosts"

Honey bees' health problems would seem to give them membership to a not-so-exclusive club of threatened and disappearing animals, or "ghosts" (Weis 2015; Tsing et al. 2017). For example, Time Magazine's cover story in August 2013 was on "a world without bees"; more colorful approaches have imagined "Colony Collapse Cuisine"—a cookbook for a world without bee-pollinated plants (Food Phreaking n.d.). More to the point, honey bees are often seen as wild creatures, impossible to fully domesticate or control, with no convincing evidence that they can recognize their keepers. As self-sufficient superorganisms, they lead lives alien to most human pursuits. More poetically, Thoreau likened their keeping to the "directing of sunbeams" (in Hyde 2002, p. 48).

Managed honey bees are generally not confined, except when they are transported. To eat and mate, they depend on diverse swaths of land too large for humans to control, up to 50,000 acres (about 80 square miles) surrounding a hive. Bringing in nectar and pollen (and sometimes things that only resemble nectar and pollen) from a wide swath of usually mixed-use land, they are unavoidably exposed to whatever is in the environment. This has most commonly been taken to mean pesticides, but more benignly, it includes whatever plants are available. And as different agricultural crops have waxed and waned in price and popularity, the forage for bees has shifted: for example, in New York State, buckwheat gave way to alfalfa, which then gave way to corn and soy, each transition less nutritious for the bees. Bees' reliance on landscapes that are not managed to protect them has left them vulnerable to environmental changes in ways that most "domesticated" livestock are not subjected to. Many of my interviewees described how their bees were imperiled by a broad range of environmental factors outside of their control.

Bees as "things"

But bees are perhaps more comparable to chickens and cows than to their native pollinator cousins. This is not a new observation; they have been called "lilliputian livestock" and their hives are emblematic of agricultural "legibility," designed for the needs and convenience of the human beekeeper (Buchmann and Nabhan 1996, p. 25; Nimmo 2015; Scott 1998, pp. 2–3). To improve their chances of survival, they are heavily managed: fed (with sugar syrup or high fructose corn syrup) when there is little forage available, medicated against infectious diseases, and given shelter that is for the most part more secure than what they can find in the wild. Furthermore, the bees themselves have been physically, behaviorally, and socially "remade" over centuries of selective breeding. They are now bigger and hairier, with thicker exoskeletons and changed digestive tracts (Kosek 2010, p. 651). Colonies tend to be less defensive and more productive than they once were.

More relevantly, today, most of the 2.5 million colonies in the US are intensively managed, "bound to industrialism [and] modern capitalist agricultural production" (Kosek 2010, p. 671; Suryanarayanan 2014). Nearly two-thirds of them are embedded in circuits of migratory pollination. These bees spend the early winter in the South, being fed (usually high fructose corn syrup) and growing in population. Most are then trucked to California in February to pollinate the growing number of almond orchards. From there, some colonies go to apple orchards in Washington State, or citrus groves in Florida, or blueberry farms in Maine. Most of these colonies are consolidated into a small number of operations with hundreds or thousands of colonies.

Thus colony numbers fluctuate based in large part on the demand (and price) for pollination contracts, honey, and wax, plus subsidies or other government or market forces, which is to say the change in the total number of colonies from year to year, both domestically and globally, is more about agricultural economics than environmental health. For instance, competition from imported honey, cheap sugar, and sugar substitutes has driven declines in the number of colonies in industrialized countries (Smith et al. 2013). Even social and political events shape the number of colonies: the collapse of the Soviet Union and the end of government financing for beekeeping halved the number of colonies in eastern Europe in the 1990s (Smith et al. 2013). Therefore, while the number of colonies in the US is roughly half of what it was in the middle of the 20th century, it was in fact the war effort that drove the numbers to an all-time high in the 1940s (because of the rationing of sugar and the use of wax for military purposes), and the numbers have been dropping ever since (vanEngelsdorp and Meixner 2010).

There is little risk, therefore, that honey bees will go extinct. Their health problems are analogous to the periodic outbreaks within industrial livestock operations such as avian or swine flu, in contrast to the implied mystery or environmental causes often described in the media. Indeed, "saving the bees" is in fact, ironically, a call to stabilize the industrial agriculture and that the very stakes of honey bee health seem so high is due to their integral role in industrial agriculture (Nimmo 2015). As one interviewee predicted:

> The beekeeping of the future is going to become much more… of a husbandry type industry than it has in the past… more like a lot of the other farming industries have become… There's going to be a lot more protection. We're going to be doing a lot more monitoring, and it's going to become much more like raising poultry or dairy farming.

He expected increased control over bees' nutrition and increased efforts to overwinter them in "environmentally controlled buildings," concluding, "we're really going to have to treat them more like a domesticated animal than a wild insect."

Bees as both?

Honey bees thus embody both halves of the twin rise of "defaunation" (declines in animal abundance and extinction) and "commodi-faunation," or the industrial-scale commodification of animals for meat and other products (Young et al. 2016; Weis 2015). They are threatened by the degradation of the landscapes that they depend on, but they are also deeply embedded in agriculture, such that they would not exist in such great numbers if it were not for that relationship. This hybridity is not a new observation: Claude Lévi-Strauss claimed that honey bees were in a "liminal zone" in between nature and culture (in Nimmo 2015, p. 186). Anna Tsing has written about the construction of bees' domesticity alongside their independence (1995).

Briefly, the perceived relationship of honey bees to other, native pollinators is instructive, insofar as it helps pin down what a honey bee is and is not. Some authors draw a contrast between them, fearful that a focus on honey bees could diminish efforts to conserve wild species, and even contribute to further declines, as the different species compete for resources and share diseases (Geldmann and González-Varo 2018). Others have found the conflation useful: the federal-level Pollinator Health Task Force report tacks between a discussion of native plants, pollinators in general, and the 15-billion dollar pollination industry, in an effort to describe (and construct) what is at stake (2015).

Being "kept" has in some ways been bad for the bees—as they give their honey up to their keepers, face the stresses of hive inspections, and more. One of my interviewees joked that the "best thing to ever happen to beekeeping was the development of [the modern hive] and the very worst thing that ever happened to bees was the invention of [the modern hive]." But kept bees are also fed and sheltered, and are likely much more numerous than they would be without such a relationship to humans. Perhaps they have domesticated us, using us for their own ends, rather like Michael Pollan's tulips and potatoes (2001).

Not-so-natural selection and fitness

Finally, one area of overlap between the two approaches to beekeeping stands out. Today, some of the most respected figures in beekeeping are looking to develop a long-term solution by breeding stocks of bees that are resistant to or tolerant of mites and other pests and diseases. One long-time honey bee behavior researcher, Tom Seeley, has introduced the idea of "Darwinian beekeeping," where genetic traits that confer better resistance to or tolerance of mites such as hygienic or grooming behavior, are promoted through a mix of genetics and limited management (2017; Neumann and Blacquière 2017). This hands-off approach entails keeping smaller colonies that are less concentrated (fewer hives per apiary) and allowing colonies to swarm (or reproduce by splitting into two, sending one half of their population out into the wild—a behavior largely suppressed by beekeepers). In short, it suggests that beekeeping should support colonies that are leaner and possibly meaner—meaning more defensive and difficult for beekeepers to handle—and not managed for maximum honey production or pollination services.

In the same vein, many beekeepers I spoke to—especially in more northern climes—are converging on the need for locally adapted populations, promoting genetics that help bees adapt to varied climates and microclimates (McNeil 2016). One interviewee described a sustainable approach as requiring "breeding and selection and… really paying attention to the genetics." She described her process of genetic testing on different bees, and how she selected certain queens.

In many ways, genetics go hand-in-hand with localization: the bees that fare well in south Florida are not the same bees that can survive a winter on the St. Lawrence Seaway. One respondent was clear that his bees had to be regionally adapted to the Northeast. But localization has its limits. I lurked during heated list-serv discussions about whether and how quickly population dynamics can reflect local environmental pressures. A state apiarist also described the need for genetics to promote behavior to make bees easier to transport, like staying calm in a moving vehicle—genetics that would be valuable for commercial beekeepers, but not hobbyists who keep their bees in their backyard.

So, despite the apparent convergence on the theory behind genetic tools, they have not actually brought the different camps together in practice: one interviewee complained that she had seen beginner beekeepers take just one piece of a successful treatment-free approach, and try to practice it without the rest of the required manipulations, meaning that they would simply buy bees and let them die of diseases: "They're not doing anything to select for better bees. The only way that that works is if the surviving ones are then the ones you breed the next year."

Conclusion

So what does it mean to "save the bees" in a world of novel ecologies and shifting baselines, where the bees themselves are a hybrid creature? Managed honey bees are emblematic of the Anthropocene—animals husbanded as gardens are tended, "monsters" shaped by scientists and bred by commercial beekeepers, and sites of struggle around the political economy of agriculture (Robbins and Moore 2013). It is not that one group of beekeepers is right or the other is wrong; rather, each has different visions of the honey bee, and in trying to save it, each is trying to save something different. One is trying to save a modern beekeeping industry that goes back more than a century, conventional agriculture with its large and distant monocultures, the livelihoods that support and depend on it, an entire way of life for much of the global North. The other group is trying to save something even bigger: an idea of "nature" that flourishes outside of human control.

Bruno Latour has written about how today's world is a bit like Frankenstein's monster (2011). In Mary Shelley's original story, the sin was not the making of the monster; it was its abandonment. Only in its abandonment did the monster became a threat. In the same way, we have (re)made our world, including the bees, and we cannot abandon them to a world for which they may no longer be fit (Lorimer and Driessen 2013). But it is not yet clear what taking responsibility for it means. And as the bees are unlikely to go extinct with or without our help, the issue is not about honey bee survival; it is about whether we think honey bees can suffer and what our responsibility is for individual colonies. More broadly, it is about our relationship to agriculture, our expectations from farmers, whether we should welcome the international trade and transport of organisms or try as much as possible to contain them, and more. There are no straightforward "solutions," just a long series of conversations about values and priorities, the slow establishment of new norms, updates to institutions like cooperative extension programs. So, despite scholars' skepticism of the human-centric focus of the Anthropocene, I conclude this article with a focus on the beekeepers themselves. They must continue to learn, and learn to live together, with what one interviewee called "beginners mind." The changes in the landscape in the decades that this particular beekeeper had kept bees have forced him to rethink what he thought he knew, what used to work. As the past is not erased, but is not a good referent anymore, it is not clear yet what sorts of knowledges will be the most useful (Maderson and Wynne-Jones 2016; Adams 2018). Let the Anthropocene be a call for clear-eyed responsibility in a changed world.

Acknowledgements This research has been supported in part by a partnership between Oxfam and the Atkinson Center for a Sustainable Future (Cornell University), as well as the Toward Sustainability Foundation. Emily Baker, Justine Lindemann, and anonymous reviewers provided helpful comments on earlier drafts.

References

Adams, Emily Caroline. 2018. How to Become a Beekeeper: Learning and Skill in Managing Honeybees. *Cultural Geographies* 25 (1): 31–47.

Bee Informed Partnership. 2018. Honey Bee Colony Losses 2017–2018: Preliminary Results. https://beeinformed.org/results/honey-bee-colony-losses-2017-2018-preliminary-results/. Accessed 15 May 2019.

Beisel, Uli, Ann Kelly, and Noémi Tousignant. 2013. Knowing Insects: Hosts, Vectors and Companions of Science. *Science as Culture* 22 (1): 1–15.

Boyd, William. 2001. Making Meat: Science, Technology, and American Poultry Production. *Technology and Culture* 42 (4): 631–664.

Buchmann, Stephen, and Gary Paul Nabhan. 1996. The Pollination Crisis. *The Sciences* 36 (4): 22–27.

Buck, Holly Jean. 2015. On the Possibilities of a Charming Anthropocene. *Annals of the Association of American Geographers* 105 (2): 369–377.

Collard, Rosemary-Claire. 2014. "Wild." *The Multispecies Salon* (blog). http://www.multispecies-salon.org/wild/. Accessed 15 May 2019.

Collard, Rosemary-Claire, Jessica Dempsey, and Juanita Sundberg. 2015. A Manifesto for Abundant Futures. *Annals of the Association of American Geographers* 105 (2): 322–330.

Collective, Human Animal Research Network Editorial (ed.). 2015. *Animals in the Anthropocene: Critical Perspectives on Non-human Futures*. Sydney: Sydney University Press.

Comaroff, Jean. 2017. Invasive Aliens: The Late-Modern Politics of Species Being. *Social Research: An International Quarterly* 84 (1): 29–52.

Crist, Eileen. 2013. On the Poverty of Our Nomenclature. *Environmental Humanities* 3 (1): 129–147.

Cronon, William. 1991. *Nature's Metropolis: Chicago and the Great West*. New York: W. W. Norton & Company.

Crutzen, Paul, and Eugene Stoermer. 2000. The Anthropocene. *Global Change Newsletter* 41: 17–18.

Davis, Mark, Matthew Chew, Richard Hobbs, Ariel Lugo, John Ewel, Geerat Vermeij, James Brown, et al. 2011. Don't Judge Species on Their Origins. *Nature* 474 (7350): 153–154.

Ellis, Rebecca. 2018. Save the Bees? Agrochemical Corporations and the Debate Over Neonicotinoids in Ontario. *Capitalism Nature Socialism*. https://doi.org/10.1080/10455752.2018.1494748.

Food Phreaking. n.d. The Center for Genomic Gastronomy. http://www.foodphreaking.com/issue0pdf/Food_Phreaking.pdf. Accessed 15 May 2019.

Geldmann, Jonas, and Juan P. González-Varo. 2018. Conserving Honey Bees Does Not Help Wildlife. *Science* 359 (6374): 392–393.

Haraway, Donna J. 2008. *When Species Meet*. Minneapolis: University of Minnesota Press.

Haraway, Donna. 2015. Anthropocene, Capitalocene, Plantationocene, Chthulucene: Making Kin. *Environmental Humanities* 6 (1): 159–165.

Hovorka, Alice. 2017. Animal Geographies II: Hybridizing. *Progress in Human Geography* 42 (3): 453–462.

Hyde, Lewis. 2002. *The Essays of Henry D. Thoreau*. New York: North Point Press.

Johnson, Elizabeth, Harlan Morehouse, Simon Dalby, Jessi Lehman, Sara Nelson, Rory Rowan, Stephanie Wakefield, and Kathryn

Yusoff. 2014. After the Anthropocene: Politics and Geographic Inquiry for a New Epoch. *Progress in Human Geography* 38 (3): 439–456.

Kirksey, S., and Stefan Helmreich. 2010. The Emergence of Multispecies Ethnography. *Cultural Anthropology* 25 (4): 545–576.

Kleinman, Daniel Lee, and Sainath Suryanarayanan. 2012. Dying Bees and the Social Production of Ignorance. *Science, Technology and Human Values* 38 (4): 492–517.

Kolbert, Elizabeth. 2014. *The Sixth Extinction: An Unnatural History*. New York: Henry Holt and Company.

Kosek, Jake. 2010. Ecologies of Empire: On the New Uses of the Honeybee. *Cultural Anthropology* 25 (4): 650–678.

Latour, Bruno. 2011. "Love Your Monsters." In *Love Your Monsters: Postenvironmentalism and the Anthropocene*, eds. Michael Shellenberger and Ted Nordhaus, 16–23. The Breakthrough Institute. http://thebreakthrough.org/index.php/journal/past-issues/issue-2/love-your-monsters. Accessed 15 May 2019.

Lezaun, Javier. 2011. Bees, Beekeepers, and Bureaucrats: Parasitism and the Politics of Transgenic Life. *Environment and Planning D: Society and Space* 29 (4): 738–756.

Liboiron, Max, Manuel Tironi, and Nerea Calvillo. 2018. Toxic Politics: Acting in a Permanently Polluted World. *Social Studies of Science* 48 (3): 331–349.

Lorimer, J. 2012. Multinatural Geographies for the Anthropocene. *Progress in Human Geography* 36 (5): 593–612.

Lorimer, Jamie, and Clemens Driessen. 2013. Wild Experiments at the Oostvaardersplassen: Rethinking Environmentalism in the Anthropocene. *Transactions of the Institute of British Geographers* 39 (2): 169–181.

Lorimer, Jamie, Chris Sandom, Paul Jepson, Chris Doughty, Maan Barua, and Keith Kirby. 2015. Rewilding: Science, Practice, and Politics. *Annual Review of Environment and Resources* 40: 39–62.

Maderson, Siobhan, and Sophie Wynne-Jones. 2016. Beekeepers' Knowledges and Participation in Pollinator Conservation Policy. *Journal of Rural Studies* 45: 88–98.

Malm, Andreas, and Alf Hornborg. 2014. The Geology of Mankind? A Critique of the Anthropocene Narrative. *The Anthropocene Review* 1 (1): 62–69.

Marris, Emma. 2011. *Rambunctious Garden: Saving Nature in a Post-wild World*. New York: Bloomsbury USA.

Mason, Philip A. 2016. *American Bee Books: An Annotated Bibliography of Books on Bees and Beekeeping, 1492 to 1992*. Boston: The Club of Odd Volumes.

McNeil, M.E.A. 2016. Bomb vs. Bond: Varroa Treatment/Nontreatment: The Twain Do Meet, and Let Us Count The Ways, Part I. *American Bee Journal*, 1137–1142.

Moore, Lisa Jean, and Mary Kosut. 2013. *Buzz: Urban Beekeeping and the Power of the Bee*. New York: New York University Press.

Mullin, Christopher, Maryann Frazier, James Frazier, Sara Ashcraft, Roger Simonds, Dennis van Engelsdorp, and Jeffery Pettis. 2010. High Levels of Miticides and Agrochemicals in North American Apiaries: Implications for Honey Bee Health. *PLoS ONE* 5 (3): e9754.

National Agricultural Statistics Service. 2017. "Honey." United States Department of Agriculture (USDA). https://www.nass.usda.gov/Surveys/Guide_to_NASS_Surveys/Bee_and_Honey/. Accessed 15 May 2019.

Neumann, Peter, and Tjeerd Blacquière. 2017. The Darwin Cure for Apiculture? Natural Selection and Managed Honeybee Health. *Evolutionary Applications* 10 (3): 226–230.

Nimmo, Richie. 2015. Apiculture in the Anthropocene: Between Posthumanism and Critical Animal Studies. In *Animals in the Anthropocene: Critical Perspectives on Non-Human Futures*, ed. Human Animal Research Network Editorial Collective, 177–200. Sydney: Sydney University Press.

Nimmo, Richie. 2013. From Monarchists to Communists: Bees in the Socio-Political Imagination. *Sociological Insect* (blog). 2013. https://sociologicalinsect.com/2013/10/10/from-monarchists-to-communists-bees-in-the-socio-political-imagination/. Accessed 15 May 2019.

Nordhaus, Hannah. 2016. *A Swarm of Controversy*. Wired. www.wired.com/2016/08/jerry-hayes-how-to-save-the-bees-monsanto/. Accessed 15 May 2019.

Phillips, Catherine. 2014. Following Beekeeping: More-than-Human Practice in Agrifood. *Journal of Rural Studies* 36: 149–159.

Philo, Chris, and Chris Wilbert. 2000. *Animal Spaces, Beastly Places*. New York: Routledge.

Pollan, Michael. 2001. *The Botany of Desire: A Plant's-Eye View of the World*. New York: Random House.

Pollinator Health Task Force. 2015. National Strategy to Promote the Health of Honey Bees and Other Pollinators. https://obamawhitehouse.archives.gov/sites/default/files/microsites/ostp/Pollinator%20Health%20Strategy%202015.pdf. Accessed 15 May 2019.

Robbins, Paul, and Sarah Moore. 2013. Ecological Anxiety Disorder: Diagnosing the Politics of the Anthropocene. *Cultural Geographies* 20 (1): 3–19.

Russell, Edmund. 2011. *Evolutionary History: Uniting History and Biology to Understand Life on Earth*. Cambridge: Cambridge University Press.

Schrepfer, Susan, and Philip Scranton (eds.). 2003. *Industrializing Organisms: Introducing Evolutionary History*. New York: Routledge.

Scott, James C. 1998. *Seeing like a State: How Certain Schemes to Improve the Human Condition Have Failed*. New Haven: Yale University Press.

Seeley, Thomas. 2017. Darwinian Beekeeping: An Evolutionary Approach to Apiculture. *American Bee Journal* 2017: 277–284.

Silverman, Chloe. 2013. How Do You Spot a Healthy Honey Bee? *Limn* 3. http://limn.it/how-do-you-spot-a-healthy-honey-bee/. Accessed 15 May 2019.

Smith, Neil. 1996. The Production of Nature. In *FutureNatural: Nature/Science/Culture*, ed. George Robertson, Melinda Mash, Lisa Tickner, Jon Bird, Barry Curtis, and Tim Putnam, 35–54. New York: Routledge.

Smith, Kristine, Elizabeth Loh, Melinda Rostal, Carlos Zambrana-Torrelio, Luciana Mendiola, and Peter Daszak. 2013. Pathogens, Pests, and Economics: Drivers of Honey Bee Colony Declines and Losses. *EcoHealth* 10 (4): 434–445.

Steinhauer, Nathalie, Kelly Kulhanek, Karina Antúnez, Hannelie Human, Panuwan Chantawannakul, and Marie-Pierre Chauzat. 2018. Drivers of Colony Losses. *Current Opinion in Insect Science* 26: 142–148.

Suryanarayanan, Sainath. 2014. On an Economic Treadmill of Agriculture: Efforts to Resolve Pollinator Decline. In *Controversies in Science & Technology: From Sustainability to Surveillance*, eds. Daniel Lee Kleinman, Karen Cloud-Hansen, Abby Kinchy, and Jo Handelsman, 259–268. Oxford: Oxford University Press.

Suryanarayanan, Sainath, and Daniel Lee Kleinman. 2012. Be(e)coming Experts: The Controversy Over Insecticides in the Honey Bee Colony Collapse Disorder. *Social Studies of Science* 43 (2): 215–240.

Suryanarayanan, Sainath, and Daniel Lee Kleinman. 2016. *Vanishing Bees: Science, Politics, and Honeybee Health*. New Brunswick: Rutgers University Press.

Tsing, Anna Lowenhaupt. 1995. Empowering Nature, Or: Some Gleanings in Bee Culture. In *Naturalizing Power: Essays in Feminist Cultural Analysis*, eds. Sylvia Junko Yanagisako and Carol Lowery Delaney, 113–144. New York: Routledge.

Tsing, Anna Lowenhaupt, Nils Bubandt, Elaine Gan, and Heather Anne Swanson (eds.). 2017. *Arts of Living on a Damaged Planet:*

Ghosts and Monsters of the Anthropocene. Minneapolis: University of Minnesota Press.
van Engelsdorp, Dennis, Jerry Hayes Jr, Robyn Underwood, and Jeffery Pettis. 2010. A Survey of Honey Bee Colony Losses in the United States, Fall 2008 to Spring 2009. *Journal of Apicultural Research* 49 (1): 7–14.
vanEngelsdorp, Dennis, and Marina Meixner. 2010. A Historical Review of Managed Honey Bee Populations in Europe and the United States and the Factors That May Affect Them. *Journal of Invertebrate Pathology* 103: S80–S95.
Watson, Kelly, and J. Anthony Stallins. 2016. Honey Bees and Colony Collapse Disorder: A Pluralistic Reframing. *Geography Compass* 10 (5): 222–236.
Watts, Michael. 2000. Afterword: Enclosure. In *Animal Spaces, Beastly Places: New Geographies of Human–Animal Relations*, ed. Chris Philo and Chris Wilbert, 291–301. New York: Routledge.
Weis, Tony. 2013. *Ecological Hoofprint: The Global Burden of Industrial Livestock*. New York: Zed Books.
Weis, Tony. 2015. *Ghosts and Things: Animal Life in the Capitalocene*. NY, July: Binghamton.
Winston, Mark. 2015. Manifesto. http://winstonhive.com/?p=504. Accessed 15 May 2019.
Young, Hillary, Douglas McCauley, Mauro Galetti, and Rodolfo Dirzo. 2016. Patterns, Causes, and Consequences of Anthropocene Defaunation. *Annual Review of Ecology Evolution and Systematics* 47 (1): 333–358.

Eleanor Andrews is a Ph.D. candidate in the department of Development Sociology at Cornell University. Her research explores the social and political dimensions of honey bee health.

The manufacturer's authorised representative in the EU is Springer
Nature Customer Service Centre GmbH, Europaplatz 3, 69115 Heidelberg,
Germany. If you have any concerns regarding our products, please
contact ProductSafety@springernature.com

Printed and bound by CPI Group (UK) Ltd, Croydon, CR0 4YY
01/10/2026
02225888-0002